Incorporating Groundwater Irrigation

WAGENINGEN UNIVERSITY
WATER RESOURCES SERIES 8

This series publishes interdisciplinary analyses of water resources issues in Asia, particularly South Asia. The volumes are mostly Ph.D. theses defended at Wageningen University, the Netherlands, under the 'Matching Technology and Institutions' Ph.D. research programme that focusses on the interdisciplinary study of tank, well and canal irrigation management in India and Nepal. The programme is implemented by the Irrigation and Water Engineering group and the Law and Governance group of Wageningen University, and is financially supported by the Ford Foundation.

WAGENINGEN UNIVERSITY

Incorporating Groundwater Irrigation

Technology Dynamics and Conjunctive Water Management in the Nepal Terai

SUMAN RIMAL GAUTAM

Orient Longman

ORIENT LONGMAN PRIVATE LIMITED

Registered Office
3-6-752 Himayatnagar, Hyderabad 500 029 (A.P.), India
Email: cogeneral@orientlongman.com

Other Offices
Bangalore, Bhopal, Bhubaneshwar, Chennai
Ernakulam, Guwahati, Hyderabad, Jaipur
Kolkata, Lucknow, New Delhi, Patna

First Published by Orient Longman 2006

ISBN 81 250 2992 3

Printed at
Graphica Printers, Hyderabad

Published by
Orient Longman Private Limited
3-6-752 Himayatnagar
Hyderabad 500 029
Email: hyd2_ollhyd@sancharnet.in

The external boundary of India as depicted in the map in this book is
neither correct nor authentic

Contents

Tables

Figures

Boxes

Pictures

Glossary

adhiya	sharecropping
bigha	0.68 hectare
bighatti	in proportion to bigha
Bikram Sambat	Official Nepali calendar. It is 56.7 years ahead of the Gregorian calendar. New year begins in mid-April. Example: year 2000 was equivalent to 2056-2057 BS.
Burmeli	Nepali people who lived in Burma (Myanmar)
chattis	thirty-six
chowkidar	gate-keeper
gaon	village
hastantaran	hand-over
hunda	contract farming
jharan	sub-surface springs and drains
katha	20kathas=1bigha=0.68hectare
khola	stream
kodalo	short-handled hoe
kulahi	canal cleaning operations
kulara	refers both to water entitlement from the main irrigation system as well as to the obligation to mobilise resources
kulo samiti	irrigation committee
kulos	canals

nala	drains
paani	water
Pradhan Pancha	elected head of a village panchayat in the Panchayat political system
puja	worship, devotional observances, devotional offerings
saacho	wooden proportioning device
samiti	committee
sana kisaan sanstha	small farmers organisation
sorha	sixteen
tole/tola	village or settlement

Abbreviations

ADBN	Agricultural Development Bank Nepal
ADO	Agriculture Development Office
AO	Association Organizer
APP	Agriculture Perspective Plan
APROSC	Agriculture Project Services Center
AsDB	Asian Development Bank
BLGWIP	Bhairahawa Lumbini Groundwater Irrigation Project
B.C.	Before Christ
B.S.	Bikram Sambat
CECI	Center Etudes des Canadien International
CGISP	Community Groundwater Irrigation Sector Project
CIMMYT	International Center for Maize and Wheat Research
CO	Community Organization
°C	Degree Centigrade
DDC	District Development Committee
DHM	Department of Hydrology and Meteorology
DIO	District Irrigation Office
DOI	Department of Irrigation
DTW	Deep Tubewell
FIWUD	Farm Irrigation and Water Utilisation Division
FMIS	Farmer Managed Irrigation System
FOD	Farmer Organization Division

GWRDB	Groundwater Resources Development Board
HMGN	His Majesty's Government of Nepal
HP	horse power
ICIMOD	International Centre for Integrated Mountain Development
IDA	International Development Association
IDRC	International Development Research Centre
IFAD	International Fund for Agricultural Development
ILC	Irrigation Line of Credit
IIMI	International Irrigation Management Institute
IWMI	International Water Management Institute
IMT	Irrigation Management Transfer
ISF	Irrigation Service Fee
ISSP	Irrigation Sector Support Project
JMA	John Mellor Associates
KW	Kilo watt
KV	Kilo Volt
KVA	Kilovolt ampere
Lit/sec	litres per second
M	metres
MCM	million cubic metres
mm	milli metres
m^3/sec	cubic metres per second
m^3/hr	cubic metres per hour
NEA	Nepal Electricity Authority
NFIWUAN	National Federation of Irrigation Water Users' Nepal
NGO	Non governmental organisation
NPC	National Planning Commission
NRs.	Nepalese Rupees
O&M	Operation and Maintenance
PDDP	Participatory District Development Programme
PRA	Participatory Rural Appraisal
REDP	Rural Energy Development Programme
SFDP	Small Farmer Development Programme

SIS	Surface irrigation system
STW	Shallow tubewell
UNDP	United Nations Development Programme
USAID	United States Agency for International Development
US$	United States dollar
VDC	Village Development Committee
W	watt
WB	World Bank
WECS	Water and Energy Commission Secretariat
WRSF	Water Resource Strategy Formulation
WUG	Water Users' Group
WUA	Water Users Association

Preface

My first experience in interdisciplinary work started early in my career, in 1989 when I was involved in field-survey in a watershed in Surkhet district in west Nepal. The team was made up of foresters, geologists, social scientist, soil scientist, agricultural engineer and support staff with background in agriculture sciences. By the end of the first few days in the field, I realized that we were all working across disciplines. I found myself taking part in the house-to-house survey (a duty assigned to the socio-economist), interviewing farmers living in the watershed, learning about trees, plants and rock types and most important, about the people themselves. A watershed management plan could not be complete if we acted alone. It was from that time that I learnt the value of inter-disciplinary work.

The next opportunity came when I was doing my Master's in the School of Civil Engineering, Asian Institute of Technology. It was at that time (1993-94) that we learnt to look beyond 'engineering'. Courses in socioeconomic aspects of irrigation were introduced within the Irrigation Engineering program at that time. Such a practice was still not common in an irrigation engineering department in those days. I also got an opportunity to conduct a social-anthropological research in a farmer-constructed-and managed-irrigation system in Nepal, for my Master's thesis.

This book is my Phd dissertation. Different events and experiences have shaped my interest to pursue this study in groundwater irrigation in surface water irrigated areas. It started in 1998 after a detailed study on shallow groundwater irrigation in all the twenty districts of the Terai. At that time, I worked as a researcher with Winrock International in Nepal and was part of the team that conducted the research. There were no tubewells in the kulo system that I studied for my master's research. However, there were many tubewells in command area of surface and deep tubewell irrigation systems when we conducted research in 1997-

98. I wanted to know more about their interactions and opportunity knocked on my door when I came to hear about the Matching Technology and Institutions Programme of the Irrigation and Water Engineering Group Department of Environment Sciences, Wageningen University, Netherlands in the year 2000. I had not thought of going back to school again. However, when I heard about the interdisciplinary approach that the research group promoted, it attracted me.

This study would not have been possible without the guidance and continued support by Professor Linden Vincent. I am deeply indebted to her for the supervision and continued encouragement that she has given me throughout the research period. The discussions that I have had with her and the critical comments and suggestions that she has given me, have been very helpful in shaping the research. I would also like to thank her for helping me find library facilities in Washington DC. This really helped me 'come back to the PhD research'. I feel greatly indebted to Dik Roth. He has been a constant source of encouragement for me throughout the research. His critical comments and suggestions have been helpful both during the field research as well as during the conceptualisation and the writing phase. I have learnt a lot from him through the long hours of discussions in 2005. I always felt more confident after that. Peter Mollinga deserves special thanks. The workshops that he organised for the MTI group and his comments and suggestions during these interactions have been very fruitful. In addition, I would also like to express my appreciation for his support in arranging the publication. I take this opportunity to thank the IWE and the Ford Foundation for facilitating and funding the study.

I would like to thank Ajaya Dixit of Nepal Water Conservation Foundation for all the help that he has given me. The suggestions that he has given me and the discussions that I had with him have been useful. He also helped me with literature while in Nepal as well as in the United States.

Two families facilitated my stay in Bhairahawa: the family of Padma B.K. and Thaneswar Pandey. Raj Kumar Baral made arrangements for me to stay with his family in Bhairahawa during 2001/02. I owe a lot to his mother Padma B.K and other members of his family specially Durga and Guddi. Thaneswar Pandey helped me throughout the fieldwork. It has been a good experience working with him. He was always a very enthusiastic and a motivated researcher and equally interested in the research. I acknowledge the help provided by his wife Prabha during my stay in Bhairahawa. A large number of people in the research area provided assistance and information crucial for this study. They were always very cooperative and enthusiastic to share their views, opinions and concerns. I owe a lot to all these individuals. The names of some individuals mentioned in the book have been changed where it was necessary to protect their privacy.

The family of Tika Ram Pandey provided me a place to stay in Tikuligarh. I would like to take this opportunity to thank them. Khim Lal Bhusal took time off from his busy schedule to accompany me in the field voluntarily. Sukhram Gupta provided help in Mahuwari. He was always willing to help despite his busy schedule as clerk in the village development committee office.

Most of the writing for this book was done in the United States and Netherlands. I would like to take this opportunity to thank Professor David Guillet of Catholic University of America, Washington D.C. and Ruth Meinzen-Dick of International Food Policy Research Institute for their help in the United States. David Guillet provided me office space and library facilities. I highly appreciate the invaluable help that he provided me. His gesture has a lot to do with setting me back on track in the research. I appreciate the help from Robert Yoder who provided me with his books and publications. They provided me valuable information on surface irrigation in the area.

I would like to thank my friend Rajendra Bir Joshi of Department of Irrigation and Niru Dahal Pandey of Department of Agriculture for all the help that they gave me during the research. Umesh Nath Parajuli and Puspa Raj Khanal also from the Department of Irrigation and Dhruba Pant of International Water Management Institute Nepal encouraged me to write the proposal for the Phd. Mr. Gaire, the hydro-geologist of BLGWIP provided me with the data on deep tubewell use. This was one of the most important sources of information for this research. Rama Shrestha, a hydro-geologist and also a friend from our professional association WPLUS, helped me with literature on groundwater irrigation in Nepal.

My years at Winrock Kathmandu have helped me shape up my thoughts and knowledge on various issues in agriculture and natural resources. It was at that time that I got an opportunity to work with noted agricultural economists like Govinda Koirala, Ram Prakash Yadav and Ganesh Thapa. The study conducted on shallow groundwater irrigation became a strong base for me during this research. At the same time I enjoyed the intellectual discussions that I had with Bikash Pandey, Ratna Sansar Shrestha and Binod Bhatta on water, energy and forestry.

The MTI group has been special. I did not have the chance to participate in all the workshops but I enjoyed the ones I did attend. I always enjoyed the discussions and friendship I had with other Phd students in IWE: Esha, Anjal, Manimohan, Amreeta, Preeta, Pranita, Conrad Zawe, Zulema and Daniel Prieto. I appreciate the interactions I had during my stay in Wageningen with Margreet Zwarteveen, Bert Bruins, Kai Wegerich, Gerrit van Vuren, Rutgerd Boelens and Jeroen Warner at the IWE group. Maria Pierce and Gerda de Fauw helped me in all the administrative works as well as to make my stay comfortable in Netherlands. A special thanks to both of them. Thanks to Bert Bruins for translating the English summary into Dutch. The Nepali students in Wageningen (in 2001 and 2005) provided a homely atmosphere. Sunil KC helped me by lending me his desktop for several months. I take this opportunity to thank him.

This book and of course, my entire studies and career would not have been complete without the strong support given to me by my parents Sushama and Gauri Nath Rimal. I dedicate this book to them. Without their encouragement, where would I be? I owe a lot to my mother who always took care of my daughter while I was away studying or busy with official work. I appreciate the help from my mother-in-law Lekha Devi Gautam for taking care of many responsibilities at home while I was busy with my work and Phd fieldwork and to my late father-in-law Balaram Upadhyay Gautam for always encouraging me for further studies. Thanks are due to all the members of the Rimal and Gautam family for all the support and love that they have given me.

I acknowledge the love and encouragement provided to me by my husband Dinesh and daughter Preetha throughout my work. They have coped with long periods of separation from time to time. However, they always provided me strength and confidence for me to continue my work. This work would not have been complete without their support.

1

Introduction

"I used to irrigate from the surface irrigation system, then I used only deep groundwater for several years. We had ignored our rights to the surface irrigation system when we got groundwater. But now our village has made arrangements to use both sources of water. Those who can afford have also installed shallow tubewells"

-Pabitra, a farmer in Madhaulia.

Groundwater development for irrigation by means of deep tubewells[1] and shallow tubewells[2] has been a key focus in rural development strategies in the southern plains of Nepal called the Terai[3]. These developments have taken place not only in many areas where agriculture was entirely rainfed but also in those that already had a history of surface irrigation management. Despite this, little is known about how groundwater is used alone or in conjunction with other sources of water for irrigation in the Terai, and what transformations in governance and productions these technology choices relate with. In this book, I examine the emergent institutions and practices that have come up for irrigation in an area that had a history in surface irrigation management, and was subject to interventions in deep as well as shallow groundwater irrigation.

The site of the study is Rupandehi district[4] in the western Terai of Nepal. A total of 182 deep tubewells were installed in this district by the Bhairahawa Lumbini Groundwater Irrigation Project (BLGWIP) from 1975 to 1999. There have been changes in the design and the process of implementation of deep tubewells through the twenty-five years history of the project. All deep tubewells that had been installed before 1992 were managed by the project. From 1992, the project handed over the deep tubewells to the water users groups that were formed for each deep tubewell management. A policy agenda on cost reduction and deregulation paved the way for turnover and transfer of deep tubewell irrigation systems, starting from the early 1990s in Nepal. This took place as part of a wider policy on irrigation management reforms. Presently, all deep tubewells are under farmer management.

Besides deep tubewells, a number of shallow tubewells are also in use in the study area. By 1999, the Agricultural Development Bank Nepal (ADBN) had installed more than 4,000 shallow tubewells in Rupandehi through its subsidy scheme. Shallow tubewells are also being installed by other non-government agencies and privately by the farmers. It is still unknown how many tubewells have been installed through these means.

Three different sites were chosen for the study area, both inside and outside the deep tubewell project boundaries. All three fall within the Tinau river basin. Before tubewells were introduced, the three sites were irrigated by different networks of surface irrigation systems (locally called *kulos*), subsurface springs (*jharan*) and drains. The latter is field-to- field drainage but is also referred to as jharan.

Research Context: Groundwater irrigation and emergent institutions

The quote at the beginning of this chapter is from Pabitra, a farmer in a village in Rupandehi. It is clear from her saying that there is an interaction between the different institutions for water management. In the following paragraphs, I present a short case of the emergence of different patterns of water use for four irrigators across the study area, to set the scene for the study.

Pabitra has a plot of land that falls in the command area of the biggest deep tubewell project in the country, the Bhairahawa Lumbini Groundwater Irrigation Project (BLGWIP). Before the deep tubewell project was implemented, the major source of irrigation for her village was the large farmer-constructed and managed irrigation system (FMIS) in the area. For several years after the deep tubewells had been installed, the project supplied free groundwater throughout the year. Pabitra and her village left the membership of the FMIS and used only deep groundwater for irrigation. The deep tubewell was then handed over to the water users' groups as part of the implementation of the policy on

FIGURE 1.1: Location of study sites along the Tinau River in Rupandehi

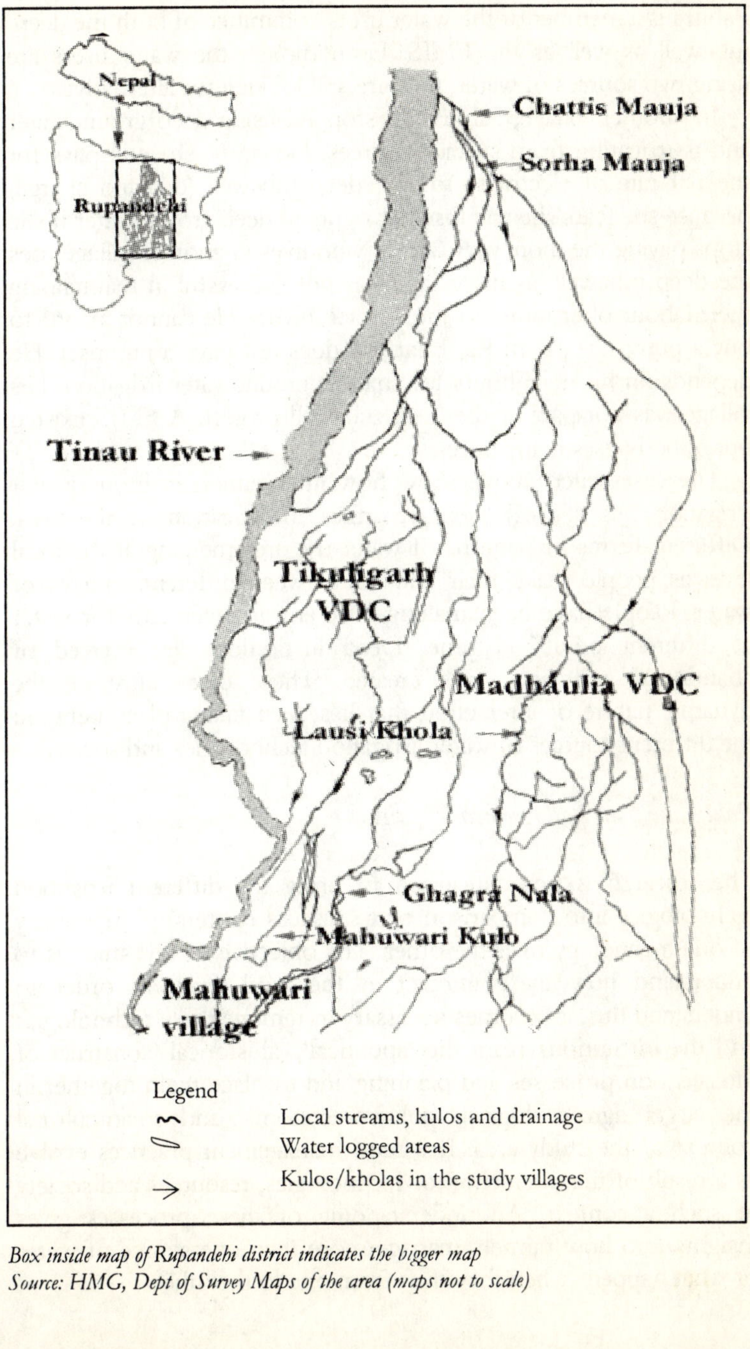

Box inside map of Rupandehi district indicates the bigger map
Source: HMG, Dept of Survey Maps of the area (maps not to scale)

irrigation management transfer. The village then made arrangements to use the surface water from the FMIS once again. Pabitra is a member of the water users committee of both the deep tubewell as well as the FMIS. Even though the water users are using two sources of water, they are still looking for alternatives.

In another village, Sabitri has stopped using deep groundwater and is irrigating from surface sources. However, she still pays for the flat rate of electricity for the deep tubewell (demand charge) because she fears she will lose her rights to deep groundwater if she stops paying the money. A farmer who lives in another village uses the deep tubewell, as his village was not successful in maintaining their labour obligations to the surface FMIS. He cannot afford to buy a pumpset[5]. Ram Raj Bhar too does not have a pumpset. He depends on his neighbours for shallow groundwater irrigation. His village was not part of the deep tubewell project. A FMIS existed but nobody uses it anymore.

The cases cited above show how intervention in groundwater irrigation has set off new dynamics in irrigation in the area. Different forms or patterns of water use are emerging at the local level as people make their choices between different sources of water. People have been making different arrangements for water at different points in time. Dead institutions are revived or abandoned and new ones created. These cases illustrate the dynamic nature of interaction that has been taking place between the different sources of water, irrigation technologies and society.

Policy questions and research objectives

The research is not concerned to bring the different irrigation technologies into comparison or to discuss matters of superiority of one technology over the other. The objective of this study is to understand how these interact at the local level. In order to understand this, it becomes necessary to remove these technologies and the institutions from the 'apolitical', 'ahistorical' construct of intervention processes and planning and to place them together in the larger agro-ecological, politico-economic and socio-cultural context of the study area. Irrigation management practices evolve as a result of the interaction of technologies, resources and society in such a context. An understanding of these processes gives insights into how farmers manage more than one source of water or what happens when they shift from one to the other.

One of the main issues constantly arising in policy making in groundwater irrigation, is that there is very low level of utilization of groundwater structures and resources in the Terai (IIMI 1991; Gautam and Shrestha 1997; Shah and Singh 2001). This has been recorded for both deep tubewells as well as shallow tubewells. Another issue that had come up was the overall disinterest of farmers to form groups around tubewells. Experience of the Agricultural Development Bank Nepal (ADBN), in implementing shallow tubewells from the early nineteen eighties till the end of the nineties, has shown that there was very little demand for group-owned shallow tubewells. The International Fund for Agricultural Development (IFAD) had to prematurely terminate its project on group tubewell installations. It took six years to install 800 tubewells (Koirala, 2001). Even though group shallow tubewells were allotted high subsidy, the total number of tubewells installed through this means was less than four percent of the total shallow tubewells installed in all the twenty districts that make up the Terai (Gautam and Shrestha 1997; Koirala 1998). The issue on low level of use has mostly been addressed from a techno-economic perspective. The constraints to group tubewells have mostly been identified in terms of difficulties such as cost sharing for repair and maintenance between farmers and social relations between neighbouring farmers sharing contiguous plots. Another objective in the study was to support more informed understanding on irrigation management related to groundwater and conjunctive use, to combat generalisations about the much discussed inability or disinterest of farmers to form groups around tubewells, and their low level of use. The other interest was so see if management of multiple sources was visible.

Policy documents and plans[6] call for the need to encourage conjunctive use of groundwater and surface water for irrigation, however, the processes of intervention in either source of water have always been isolated from each other. Therefore, the issues and concerns that have come up for irrigation, even in areas where farmers use more than one source of water, have also emerged and been treated in the same isolated manner.

Intervention processes are underway in the Terai, both in groundwater irrigation as well as in surface irrigation. Besides the Department of Irrigation, some non-governmental organisations and bi-lateral projects are involved in irrigation. The non-government sector has been mostly involved in the dissemination of smaller irrigation technologies like shallow tubewells, treadle pumps and sprinkler irrigation as well as in the rehabilitation of small surface irrigation systems. All the intervening agencies work under the banner of a 'demand-based participatory approach'. However, each agency works in isolation from the other. Each has its own technology with a particular institution "crafted" around it. These technologies and the associated institutions have been introduced into the local setting by means of different programmes and intervention. Each programme of intervention is dealt with by the agency responsible for it. The fact that other technologies and sources of water exist is not denied. However, the target-oriented nature of the intervention programmes tends to largely deny the existence of other processes in the same locality. This attitude by the implementers of 'turning a blind eye' paves the way for more programmes. Besides, the problems in implementation of each individual intervention processes are also identified in a similar isolated way, and more and more programmes are brought in into the local context. So, how do the farmers make their choices of technologies and institutions for irrigation? And how does this affect the technological and organizational performances of each intervention? These have been some of the several problem contexts guiding this research.

The need for the study also arises from several concerns that are related to irrigation development in general and groundwater irrigation in particular. Several questions can be raised from the issues related to the stated irrigation policies and modes of intervention. Nepal Irrigation Policy (1992, 2004) stresses the need to encourage participatory approaches to irrigation management. In order to achieve this objective, several activities are being carried out. One of them involves the transfer of management of irrigation projects to the water users group. The other is the policy that gives the farmers the freedom to choose or create irrigation systems of their choice. These are done under the heading of 'demand-based' 'participatory approaches'. In these processes, a group of farmer can get together and request for a surface irrigation system, request for rehabilitation of existing farmer-managed irrigation system (FMIS) or a tubewell for irrigation. Farmers can also install shallow tubewells through different non-government agencies and also privately at their own cost. There are some legal regulations governing legitimate access to water resources. The Water Resources Act Nepal of 1992 outlines these rights. These rights define how the farmers can procure water resources. Activities like rehabilitation of FMISs or creation of new infrastructure for surface irrigation involve many interactions by the farmer groups with other irrigation canals that either share a common source with them or will be affected by the new construction in one way or the other. However, groundwater tubewells can be sited anywhere that the farmers want. There are no restrictions on this. This is especially true in case of shallow tubewells that can be owned by an individual farmer. A farmer, therefore, could site them in his plot, which could be situated within a command area of a surface irrigation project or a deep tubewell project. One of the matters of interest in this research was also to understand how these policies that aim to achieve more 'farmer participation' in areas of irrigation management translates in areas of complex water resources.

All these questions are even more relevant, when one takes into account the present focus of irrigation development that is geared towards providing increased groundwater services to the farmers. The Agriculture Perspective Plan (APP) of 1995 brought groundwater irrigation into the mainstream of irrigation development. The twenty-year plan was formulated in 1995 and is under implementation since 1997. It envisaged attaining an increase of the agricultural growth rate by two percentage points, from three to five per cent per annum. This means a six-fold increase in agricultural growth output per capita. This increase should be realised by focusing on different key inputs (APROSC/ JMA 1995), one of which is groundwater irrigation. It emphasized groundwater irrigation as 'one of the major inputs that can help boost up agricultural production in a shorter time period and tackle food security concerns'. The argument is based on the fact that groundwater has the capability of providing year-round water unlike surface sources, and that it offers more control over the water resources by the farmers. It bases its argument on the large estimate of groundwater available in the Terai. It has been estimated that 3.8 percent of the total of 233 billion cubic meters of renewable water resources in the country, is in the form of groundwater reserve in the Terai (WRSF/WB 1997). A large part of it has still not been used. The plan emphasised shallow tubewell irrigation. This technology, which was once considered a rather small and insignificant technology by agencies involved in the construction of large-scale irrigation projects in the Terai, became the most important technology to be disseminated to farmers.

Massive investments had already been done in tubewells on the Indian side of the Indo-Gangetic plain before the end of the 1980s. According to O'Mara (1998) such investments posed a challenge to resource management, but were also a 'precondition of efficient conjunctive use of surface and groundwater'. Investments in tubewells started in the Nepal Terai in the beginning of the nineteen-eighties, later than on the Indian side. It has to be seen yet how proliferation of wells affects the existing institutions around water and how the different arrangements and organization around groundwater are also affected. Conjunctive use has also always been a recommended practice. Most study reports and policy[7] documents encourage using both groundwater and surface water for irrigation in order to increase efficiency in the use of water resources (Agriculture Perspective Plan 1995; Irrigation Policy 2004; Gautam and Shrestha 1997; Koirala 1998; Koirala and Gautam 1998). But it is not clear how this is to go about. Several scholars have also developed conjunctive use models for different river basins and irrigation systems in the Terai[8]. These have, however, been limited to academic circles and not put into actual practice. The term 'conjunctive use' has been defined in various ways. It has been defined as: 'combined use of surface and ground water systems to optimize resource use and minimize adverse effects of using a single source' (Utah Water Resource 2005). IDRC (2005) defines it as, use of both surface and underground water for a single purpose, most commonly irrigation. According to Vincent and Dempsey (1991) conjunctive use is the combined and integrated management of surface and groundwater for optimal productive and allocative efficiency. Hoogesteger (2005:20) in his study on drought management strategy in Zayandeh Rud river basin in Iran refers to conjunctive use as a 'situation within an irrigation system, where farmers have access to-and use canal- and groundwater for irrigation of their fields'. He refers to conjunctive water management as an active management of both surface and groundwater by an institution. In this study, conjunctive use refers to a situation where farmers choose more than one source of water, not necessarily only groundwater or water from surface irrigation systems, but also the use of drains and other springs, in order to fulfil their irrigation water requirements. I examine how farmers in the study area make their choice between more than one source of water and types of institutions that come up around the different technologies and sources of water for conjunctive water management.

Several studies have been conducted in FMISs in the hills and plains of Nepal. The first pioneering studies on kulos were documentation and examination of organisation of surface irrigation by farmers (Martin 1986; Yoder 1986; Pradhan 1989). Other studies that subsequently followed, dealt with such issues as intervention in farmer-constructed and managed irrigation systems, comparative studies on government managed and farmer managed irrigation system, performance measurement of FMISs, water rights, farmer-managed irrigation systems and irrigation technology; and gender issues (Shivakoti 1992; Pant 2000; Lam 1998; Gautam 1994; Pradhan 1990; Pradhan et al. 2000; Shukla et al. 1996; Parajuli 1999; Zwarteveen and Neupane 1996).

Even though deep tubewells and shallow tubewells have been in use for more than two decades, the number of studies on groundwater irrigation and on the use of multiple sources of water in the Nepal Terai is very low. Most of the studies that have been conducted in groundwater irrigation have either been done to measure performance of installed tubewells or to identify the constraints to expansion of groundwater irrigation. They have been conducted with a more techno-economic perspective to provide policy level solutions in groundwater irrigation. They have, therefore, identified various technological and socio-economic constraints as contributing to the emergence of these issues (IIMI 1991; Gautam and Shrestha 1997; Koirala 1998; Koirala and Gautam 1998). H. N.Bhandari (1999) conducted an academic study on the economics of groundwater irrigation rice-based systems in the Nepal Terai. Other studies have been mostly confined to project documents and process documentation by consultants to the groundwater project. Study on the process of management transfer of deep tubewells of the BLGWIP has been documented by Olin (1994). There have not been studies based on individual deep tubewells. Myint (1999), in his paper on the experience of the World Bank with groundwater irrigation in Nepal, reports that deep tubewells were regarded as government property by the farmers. He further writes that BLGWIP was one of the most successful projects to be implemented through Bank assistance in the Terai after modifications in design and incorporations of the participatory approach, and that it was an effective model for groundwater development.

Other studies specific to Rupandehi have been those by Gyawali and Dixit (2000), who have documented the existence of multiple institutions for water management in the Tinau River basin. The Chattis Mauja irrigation system has been a subject of study by many researchers[9] and has often been cited as an exemplary case in community management. The major source of surface irrigation in the study area is also the Tinau River and the Sorha-Chattis Mauja irrigation system is the largest farmer managed irrigation system inside Rupandehi. The territory of these irrigation systems were also parts of the BLGWIP. One of the studies that were conducted in the study area is that by Shrestha and Sharma in 1986. In this study, the researchers conducted a comparative study in several villages that were irrigated by both deep tubewells as well as Chattis Mauja irrigation system. The study was conducted before 1986, when the project still worked in full supply-oriented mode. In this study, the researchers write that farmers were not willing to pay water tax even when they realised that the amount was not very big. According to these researchers, the farmers were of the belief, that once they took part in the act of payment, they would be given the responsibility of managing the tubewells.

The other development-policy related realm of study in groundwater irrigation elsewhere in Asia have focused on groundwater use especially in arid areas of India like Gujarat and in Pakistan (Bhatia 1992; Moench 1994; Shah 1993; Meinzen-Dick 1996; Dubash 2002; Prakash 2005). Studies done on groundwater irrigation in neighbouring Indian states in the Indo-Gangetic plains include those by Clay (1972), Pant and Rai (1985), Pant (2004), Ballabh et. al (2003) and Kishore (2004).

The study by Clay (1972) focuses on innovation, inequality and rural planning and economics of tubewell irrigation in the Kosi region, in Bihar in India. Others mostly deal with the groundwater irrigation and agrarian question in these states. Pant and Rai (1985) study the problems and issues related to small farmers in irrigation in eastern Uttar Pradesh and North Bihar and describe the evolution of successful experiments in community tubewells in the area. The recent study by Pant (2004) was done as a resurvey in some of the same villages in which a study had been conducted in 1985. He investigated the trends in groundwater irrigation in eastern and western Uttar Pradesh since that period of time, in terms of: changes in socio-economic status of the farmers, role of groundwater, agricultural productivity and relations between class - caste and ownership of agricultural implements. His study reported significant changes in the twenty years gap. Some of his findings relevant to this study were those on the role of public irrigation. He found that there was a decreasing role of public irrigation but there was a variation between the eastern and the western parts of the state. There was greater dependence on canals than on state-owned tubewells in the west, while farmers in the east depended more on state tubewells.

The studies in the Terai have been done with a focus only on groundwater. Most of the issues and problems that have been identified in groundwater irrigation have been addressed without placing groundwater within the larger hydrological, agro-ecological, technological, institutional and political environment in which it functions. There is a need to look into the issues and debates in groundwater by placing it in this larger context. This is all the more necessary in areas where farmers have access to more than one source of water for irrigation. There have not been any studies in the Terai that have examined the evolution of institutions for irrigation management in an area with a history of surface irrigation that has been subjected to intervention processes in groundwater irrigation.

Conceptualising Irrigation in Complex Water Resource Situations

This study is based in an area with multiple sources of water. Before the tubewells were installed, the farmers had knowledge of the management of surface sources as well as sub-surface springs. The surface irrigation networks that had been constructed by the farmers had already undergone physical as well as organizational transformations by the time the groundwater tubewells were installed. The farmers had already developed their own concepts of rights, rules and norms and ways of managing the surface water sources and drains.

Irrigation development is an ongoing process. The study area can be visualised as an area where several processes of interventions in irrigation have been going on for three decades. Parallel sets of activities were going on in deep tubewell irrigation in the study area before 1999. The BLGWIP was involved in handing over deep tubewells that had been installed, while at the same time it was also arranging other farmers without deep groundwater for new sets of deep tubewells, through its 'demand-based' approach that had been introduced in its final stages of the project. Installation of shallow tubewells is still ongoing process throughout the district. Besides these interventions in groundwater, another intervention of relevance to the study area is that of the rehabilitation programmes for farmer managed surface irrigation systems that are carried out through the irrigation sector reform programme.

In this study, I move away from the assumptions of intervention policies that expect some form of linearity in terms of interventions and outcomes. According to Long and van der Ploeg (1989) and Long (2001), "the separation of 'policy', 'implementation' and 'outcomes' is a gross over-simplification of a much more complicated set of processes which involves the reinterpretation of transformation of policy during the implementation process itself, such that there is no straight line from policy to outcomes".

In order to understand irrigation practices as they take place in this area of complex water resources, I make use of several conceptual and theoretical insights: irrigation as a sociotechnical phenomenon and legal complexity. In addition, I integrate other complementary concepts from innovation studies.

Sociotechnical complexes

In this study I conceptualise the different technological interventions as 'sociotechnical complexes within a water resource system' (Vincent 1997), that shapes and are shaped through interactions with the hydrology of the water resource system. Understanding the technology in use helps unravel the complexity of irrigation systems. Technology, according to Benton (1992), is a mediation between society and natural resources (Knegt and Vincent 2001, Vincent 1997, Mollinga 1984 in Vincent 2001, Vincent 2005). Farmers use different technologies to mediate water supply within the water resource system and work out innovative ways to gain control over different sources of water. The water use complexes and practices that emerge in areas with multiple sources of water can be conceptualised as social constructs that come up when farmers make their choices between different sources of water at different points in time. They develop through the interaction of the different characteristics of the various technologies, the various sources of water that are used, and society; thus they are sociotechnical in nature. The process of evolution of these complexes is dynamic and they change their forms in different periods of time, until the farmers perceive a sense of security. They construct technological performance from multiple sources closest to their priorities and negotiating capabilities and that are affordable and profitable for them. The irrigation practices emerge out of interactions between different 'sociotechnical systems' but are themselves sociotechnical in nature. Farmers either choose one or use more then one source of water conjunctively for irrigation and adjust it to the natural water cycle.

The sociotechnical perspective to the analysis of irrigation practices as developed by Mollinga (1998) sees irrigation practices as coming up from interactions between technical, organisational, socio-economic and political dimensions of water control embedded in the agro-ecological system, the agrarian structure, the state and institutions of society. These in turn shape the inter-relationships of water, technology and forms of organisation (Mollinga 1998, 2003a). Roth (2003a) focuses on the relations between technology (as material infrastructure), organisational and normative dimensions of irrigation. According to him, irrigation systems can be analysed as 'intricate complexes of physical-technical, organisational and normative-legal dimensions of water control that develop in a wider agro-ecological, politico-economic and socio-cultural context' (Roth 2003a: 33).

Technical, normative and organisational interdependence between systems

According to Boelens (1998), in order for irrigation system to function, there has to be some sort of stability between the different dimensions of it: the infrastructure, normative and organisational system. Every time a new technology is introduced, it structures the way users have to use it, because it is designed in a particular way. It has, what is called its own 'social requirements for use' (Bijker 1987, Mollinga and Mooij 1989, 1989, Artifacto 1990, Boelens and Temmink 1990, Van der Ploeg 1991 cited in Boelens 1998, Mollinga 2003). It introduces completely new sets of 'technical, normative and organisational systems' into the local context. Each irrigation technology is introduced with the object of improving the water supply and it comes along with its own definition on how it should be used. Therefore, irrigation practices will evolve through the interaction between the different technological, organisational and normative characteristics of the different technologies introduced.

As more and more implementation programmes in irrigation are executed, the process of adding on to or disrupting the existing state of conformity between the 'technical, normative and organisational' dimensions of systems goes on. According to Boelens (1998), change in one dimension affects the other. In this study, change in one dimension within one system changes not only the other dimensions within it but also affects the dimensions of other forms of irrigation in the vicinity. This is because people are interacting with several sources of water. This change can be an effect of 'imposed' external intervention or can be brought about by the different actors themselves when they try to negotiate a more 'synchronised' (Boelens 1998) form of irrigation from multiple sources of water. A whole process of stabilisation, 'shaping', and 'becoming' goes on. How new technologies and interventions interact with the existing system depends on the 'state of affairs' that existed in the prevailing system. It also depends on the way people can choose between sources of water given opportunities and constraints of available water sources and intervention processes.

'Development arena': a metaphor to understand the heterogeneous nature of interactions

In order to examine or visualise the heterogeneous nature of interactions between the different sources of water, technologies and people and the way they mediate their water supply in the study area, I use the concept of 'development arena' as suggested by Jorgensen and Sorensen (1999) in their study of cognitive spaces for research and developement The notion of 'development arena' is a metaphor and it serves as a frame and mental space for discussion that focuses on different dissimilar processes that are linked to each other. According to Jorgensen and Sorensen (1999), 'it is a cognitive space where political, social and technical performances related to a specific technological problem takes place'. It helps to understand the linkages between the different actors, artefacts, the different locations for action and the processes involved in choosing, shaping, and becoming in technological change. 'It is a framework to assemble the locations and processes involved in innovation' (ibid.: 425). The concept of a development arena allows a focus on technology; but is also complementary with the idea of a social arena[10]

Practice, agency and power

In order to analyse the strategic behaviour of the farmers, and to examine the existing irrigation I make use of the concept of practice. The concept of practice helps to understand irrigation as it takes place in reality. It helps to look at the relation between intervention and reality critically. Mollinga (1998: 20), citing Giddens (1976) explains practices as what people do, in a structured and structuring fashion. According to him, social interaction is a type of practice in which people encounter each other (Mollinga 1998, 2003). In this study, the major intervention in irrigation has been in groundwater irrigation. I examine the reality of water use practices, rather than making assumptions on how it should be. I examine how farmers work out different strategies to incorporate and adjust to different processes of intervention in groundwater.

The concept of human agency is used to understand and analyse the behaviour of different actors. The concept of human agency implies that social actors have 'knowledgeability' and 'capability' and that they work out ways of coping even when subject to different (physical, normative or politico-economic) constraints and uncertainties (Long 2001). In this study, I analyse how different actors choose, reject, strategise, manipulate and adjust to work out ways for irrigation that are most effective for local production options and make use of different technical, organisational, normative/legal options in order to negotiate their water rights. Persons or networks of people also have agency (Long 2000). Farmers form different formal and informal networks to use them in strategic bargaining. Networks can also be said to have agency.

The concept of 'politics of production' from Burawoy (1985) has been used to understand the process by means of which farmers strategise and work out different production options. This ultimately affects the way they make their choices between different sources of water. According to Burawoy (1985), organization of work (labour process) is regulated by the 'political and ideological apparatuses of production'. In addition it has an economic element. The 'political and ideological apparatuses of production' help reproduce the relations of labour process through the regulation of struggles. He terms these struggles as the 'politics of production'. This gives attention to both the internal or micropolitics as well as factors external to production which affect it.

The actions and behaviour of the farmers have to be analysed by looking at how these actions are embedded in wider structures. This includes the process whereby they mediate and negotiate, make decisions while claiming rights to different sources of water and react in times of uncertainty, The behaviour of different actors in irrigation or interaction between people and the resources is embedded in the agro-ecological, technological, political, economic, social, cultural and historical context (Mosse 1997, Mollinga 1998, McCay 2002, Roth 2003). According to McCay (2002), the analysis of the interaction between people and 'common pool resources' has to be analysed beyond only looking at decision making 'calculi' of individuals. She states that it is necessary to know their backgrounds, the social entities that represent them or they help reproduce, their histories, values and significance.

Water use and management practices develop in an ongoing process whereby farmers try to gain control over one or more sources of water for irrigation. Mollinga (1998) defines water control as politically contested resource use. According to him, water control as a concept brings together the different dimensions of irrigation together: the technical, organizational, socio-economic and political control. Therefore control implies power. He uses these concepts to analyse processes within an irrigation system. However, this concept of control can be extended to understand contestation amongst and between different sources of water in a water resource system. Different types of power come into play in the process. Dowding (1996) provides two concepts of power which he terms as 'power to' and 'power over'. The first concept refers to the outcome power or the ability to bring about or help bring about outcomes; while the second one is 'social power'.

Interventions have the power to structure the way farmers have to behave by introducing a specific technological, organizational and legal elements. Power is also a relational concept and having power does not entail that others are without it: there is no zero-sum game (Long 2001). Power and control relationships help to analyse the capacity of the farmers to make use of, transform different processes of intervention, work out different relations in productions and come up with innovative solutions for irrigation. Besides, the concept of power also helps to understand the interaction that takes place between different groups of people involved in making decisions for the use of surface water and groundwater. In addition, it can be used to understand the conflicts between different groups of people who compete for water, between different political leaders who play an active role in negotiating water rights and also the power differences between different rules and regulations regarding water.

Legal Pluralism, Institutions and Property Rights

The study focuses on the relationship between the different technologies and organizational and normative dimensions of irrigation that comes up in interaction between different sociotechnical complexes. The concepts of practice, agency, process and control help to form the linkages between the different dimensions. Another element that requires attention is the normative legal dimension of water control. 'Legal complexity' is a suitable concept for the study of law and human behaviour. 'Legal pluralism' or 'legal complexity' refers to the existence and interaction of different legal orders in the same socio-political space (Benda-Beckmann 1997, 2002). 'It allows for the possibility of existence of more than one legal system in society, and unpredictable forms of interaction and hybridization between them' (Benda-Beckmann 1997, 2001, 2002). The diversity of rule-making is looked at from the perspective of legal pluralism.

The concept of institution in this thesis refers to arrangements that emanate between different farmers for irrigation, from their daily interactions and practices. This is an outcome not only of their social, political and economic relationships but also their interactions with external intervention processes and the physical conditions (which includes location, nature of resources and technology). An institution, according to Uphoff (1986), is a combination of roles, rules, procedures, a practice and a system of relations. McCay (2002) states that the concept of institution should also include patterned behaviour besides rules and she locates institutions as major features of the cultural, cognitive, and ecological realms within which acting and decision-making individuals and social groups are embedded. Therefore, in order to understand the interactions between people, resource and technology it is therefore necessary to look not just at the decision-making of individuals but also more specifically on the background of the different actors and what they do for managing the water resources.

In this study, I also examine the relations between processes of creation of property rights to different water resources and irrigation management practices that evolve around these resources. Property rights cannot be referred to simply in terms of ownership. It also includes different types of 'bundles of rights' which the farmers can draw from. It includes several types of rights and responsibilities to different forms and uses of resources. Generally, a distinction can be made between rights to use the resource and decision-making rights to regulate and control the resource use (von Benda-Beckmann and Spiertz 1997, Schlager and Ostrom 1992).

Research Question

Conceptualising different technological interventions as 'sociotechnical complexes within a water resource system' helps to bring the otherwise isolated interventions together for analysis. As the major technological intervention studied is in groundwater, the research question has been posed as: What practices have evolved for accessing water for irrigation in the groundwater intervention areas of Rupandehi in western Terai of Nepal where there was a pre-history of surface irrigation and by what processes do farmers gain control over surface water and groundwater for irrigation? In order to understand this, I seek to understand how different water sources give constraints and opportunities to different irrigators in securing water supplies; how farmers have incorporated different processes of irrigation intervention; how and why certain actors have played a role in securing control over the different sources of water; and examine how possibilities of profitable agriculture interacts with social power to shape choices of water use.

Research Sites and Methodology

Groundwater irrigation is practiced in all twenty districts of the Terai in Nepal. I have chosen Rupandehi District to be the site of the study for several reasons. Both deep tubewells and shallow tubewells have been installed extensively in this district. It accounts for the largest area developed through intervention in deep groundwater irrigation through the BLGWIP. A total of 20000 hectares was developed by the project (BLGWIP 1999)[11].

This study was conducted in three different sites inside the district. The methodology was built around case studies selected after some preliminary surveys in which hydrological complexity was a first challenge to defining approach by village or technology. After doing a preliminary survey of three Village Development Committees (VDCs) and two villages inside the district, I decided to settle for detailed study of two VDCs in the deep tubewell irrigated area and a village outside the 'project area' where farmers irrigate from shallow tubewells. A VDC is the lowest level of government body for planning. It is made up of nine wards. Each ward is made up of one or more then one settlement. This depends on the number of households and can range from 10-15 to 100-120, depending on the geographical region. The two VDCs chosen within the deep tubewell zone were Tikuligarh and Madhaulia. They lie adjacent to each other. Administratively they stand as separate bodies but physically they form contiguous areas. The third site chosen was the village of Mahuwari. It lies in the southern part of the district and it connected to the two sites via a gravel road that links it to the Lumbini highway. This highway connects with the Siddhartha highway in Bhairahawa. All three sites fall within the Tinau River basin.

I had two people who assisted me in my fieldwork. One of them worked only in the initial phase of the research and was involved in preliminary interviews. She moved to another district and I could no longer take her help. The second had a master's degree in agriculture and animal science. He assisted me throughout my fieldwork in all three study sites.

We first made a preliminary survey of Tikuligarh and Madhaulia. The preliminary survey of the area started with a 'walkthrough' across the VDCs in order to get an idea of the way the deep tubewells were located. This was later followed by a 'bicycle ride-through' because of the large area the VDCs encompass. There are altogether seventeen deep tubewells in the two VDCs. Each deep tubewell was designed to irrigate around 120 ha. The technical details of the deep tubewells were obtained from the project report of the BLGWIP. The two areas contain the oldest sets of tubewells installed by the project. In addition, Tikuligarh has four deep tubewells that became operational only from 1999. These deep tubewells are also different in design from the rest of the deep tubewells. The choice of Tikuligarh, therefore, provided a range of different deep tubewell technologies, with different pump size, yield and implementation approaches. The first initial estimate of the number and location of shallow tubewells in the three sites was taken from the database of the well certification records of the Agriculture Development Bank[12]. The records were updated in the field visits through field counts. The data for the period after 1997 were updated by contacting the local ADBN office in Bhairahawa and the Groundwater Project Office in Butwal. Records of use of deep tubewells are kept by the water users' groups of the respective tubewells. These records were requested from the water users' group and put into a database to analyse the extent of use of the deep tubewells. The BLGWIP also keeps records of well use. This was also used as a reference.

The farmers have been using deep groundwater from the early 1980s. It was obvious from the first preliminary survey of the field that the farmers used other sources of water in addition to deep groundwater. In order to understand this, we conducted two more steps in the preliminary survey. The first one was a survey of 85 randomly selected farmers inside the two VDCs. The results of the survey showed that most of the farmers made use of more than one source of water for irrigation in this area where deep tubewells had been designed to be the major source of irrigation. It was also seen that in some cases farmers used deep groundwater, shallow groundwater and kulo water. This survey helped to give an overall picture of the socioeconomic status of the farmers who used different combinations of water used for irrigation, the extent of use of the source and the choice of crops and cropping patterns. This survey combined a semi-structured questionnaire and long discussions with the farmers to get an overall situation of the history of use of water resource in the area. Even though the survey was still in its preliminary stage, it helped me clear my doubts on situation in the study area in terms of the extent of diversity in water use. The next step was to get a detailed account of the linkages between different sources of water used.

An intricate relationship between different kulos, jharans, deep tubewells and shallow tubewells started emerging. Unavailability of a detailed map of all water resources inside the VDCs made the task very difficult. Moreover, the hydrology is so complex that it was difficult to trace a discernable network in the surface and the jharan sources. The drains are naturally occurring channels that flow up when upstream villages irrigate, and the sub-surface flow gets augmented with the field-to-field drainage, so that they come up as surface flow.

It was easier to work out the networks when the surface irrigation canals inside the villages were part of a larger irrigation system. However, there were periods in time when these villages were not recorded as being part of the surface irrigation systems. The irrigation systems did not denote these villages as their designated 'command area'. These VDCs officially are 'parts of the BLGWIP'. The project boundary of the BLGWIP and the irrigated area of the Sorha and Chattis Mauja overlap in several VDCs.

The most difficult part was to find the linkages between the jharans. The difficulty was added to by the way the network of the deep tubewells intercepted the surface irrigation kulos and the jharans. Each deep tubewell was designed as a single unit with no connection with each other. These canals criss-cross across the landscape in such a way that sometimes a part of the 'command area' of the deep tubewell is crossed by one kulo and another part by a different kulo. In other cases, farmers of two deep tubewell areas share a common kulo. Therefore the next step was to discern this complexity.

For this we conducted what can be called an 'add-on' survey to the preliminary survey. It was done through a combination of interviews with key informants, field visits, semi-structured interviews, open-ended, non-standardised interviews with the farmers. It also included an exercise where we identified the location of the deep tubewells, the layout of the kulos and the drains and the location of shallow tubewells in the process.

Information on the history around water use was then collected through key informants from across the VDC. These were selected mostly based on their age and involvement in the deep tubewell and surface irrigation. Oral history on the transformation in water use in the villages was recorded through discussions with key informants mostly the local Tharu population and the first hill migrants into the village. Many times, this involved contacting people from other VDCs also. These were people who had been actively involved in surface irrigation and had been living in the area before the start of the BLGWIP.

The next step in the research was the selection of cases for detailed study. The fact that farmers had been making use of different sources of water was verified through the surveys and studies. The pattern in which they were organized was quite different from the way the deep tubewell project had 'designed' it to be. After gaining knowledge on the diversity of practices, I chose three cases inside Tikuligarh VDC as a basis for detailed study. These included areas where farmers made a combination of 1) deep groundwater and surface kulo 2) deep groundwater, jharan and field-field drains and 3) deep groundwater and shallow groundwater. The design of the deep tubewells in the first and second was similar, while it was different for the third case. In order to make a comparison with the first case of deep tubewell and kulo in Tikuligarh, I chose a similar case in Madhaulia VDC for detailed analysis.

In both cases I have started with the VDCs as the main point of entry. This approach has been chosen for several reasons: first of all the hydrological boundaries of both kulos and drains cross the boundaries of a single village. Similarly, a single deep tubewell sometimes irrigates more than one village. The administrative, social and hydrological boundaries are blurred when people start shifting between sources of water. Secondly, all deep tubewells are being managed by the farmers since 1992. After the deep tubewells were transferred, they automatically became the property of the water users' groups and indirectly became a VDC infrastructure. Therefore all deep tubewells that belonged to the BLGWIP became part of the VDC in which they were installed. The VDC is the lowest level local government body for planning. In order to capture the interactions and influence between the ways the different sources or complexes of water use are managed, it is necessary to place it in the wider political structures of the locality and the region. An approach from the VDC perspective also helped to bring out the linkages between socio-cultural characteristics in the area, the development of settlements, history of resource use and power structures. I have taken a village as a unit for detailed study in case of the third site. Interactions around shallow tubewells are more individualistic. A village approach made it more feasible to capture the relations between people, technology and resources.

A socio-anthropological approach was taken in order to understand village society and the changes around it. Yin (1984: 23) defines the case study research method as an empirical inquiry that investigates a contemporary phenomenon within its real-life context; when the boundaries between phenomenon and context are not clearly evident; and in which multiple sources of evidence are used. The tools and techniques used were in-depth interviews, case studies and participant observation. For in-depth interviews, I followed unstructured, open-ended, and non-standardised methods. These methods were used both in the preliminary stage of the survey as well as for the focus case studies.

The focus during in depth research the case studies sites strove to understand the historical changes that each specific site had undergone in terms of water use, current water use practices; and why farmers have been making their choice of sources of water individually and as a group. This involved understanding farmer strategies. In order to do this, I have used a combination of methods. One was participant observation, watching and discussing how farmers actually made their arrangements. A case study can be characterised as a detailed examination of an event (or a series of related events) which the analyst believes exhibits some general theoretical principle.... 'The focus of a case study may be a single individual as in the life-history approach, or it may be a set of actors engaged in a sequence of activities' (Mitchell 1983 as quoted in Magadlela 2000: 21). Accordingly, I have identified the main actors in the area who were responsible for bringing about or were the reasons for bringing in changes in water use in the area. The identification of such actors had to be gained through key informants as well as the general men and women in the village who had knowledge about such issues and through observations, interactions and detailed discussions with these actors themselves. Sometimes in such situations it becomes very difficult to get to 'both sides of the coin' or different 'parties' or 'groups' of people. There might also be a situation where the researcher suddenly finds himself or herself totally in interaction with the same network of people. One has to remember that in politically sensitive areas and situations people refer you from one person to another. In many cases they are discreetly referring you to people of the same political background or group defined by some similar purpose or motive. So sometimes the information that is collected might not shed real light on the issue in question. This sort of tendency was very obvious in one of my study sites. People sharing different political ideology were sort of not 'counted' by those who held more power at that instance. In that VDC, one political party held a majority while in the other, the power balance between two opposition parties was more balanced.

The management of deep tubewells and surface irrigation

involves a host of rules, regulations and local laws. The deep tubewell water users group and the water users association of the kulo (*Kulo Samiti*) both exist as formal organizations. The kulos in the study area were either part of the Sorha Mauja or the Chattis Mauja, or were independent irrigation systems. In order to understand the relation between these village level institutions and the larger networks, the written constitutions of the respective kulos was studied in detail. Discussions were also held with some members of the main committee of these organizations. The minutes of the meeting of the deep tubewells were also read when the deep tubewell water user groups were willing to share it. Rules imparted by the project for the water users groups during the process of handover were also discussed with the water users groups of the deep tubewells. I have also had discussions with some of the officials who were responsible from the project side in the process of handover of the deep tubewells. Similarly discussions were held with drilling mechanics, workshop owners, pumpset dealers and staff of non-governmental agencies involved in shallow tubewell promotion.

To understand more about the performance of technologies in relation to farmer's coping strategies, water measurements were done in selected farmers' fields in Mahuwari. Discharge measurements were carried out in 26 farmers' fields out by means of the trajectory method. The details of the cost of operation of shallow tubewells were computed for 10 out of 26 farmers in Mahuwari. The cost of production for paddy and wheat and cost sharing between landowners and sharecroppers were calculated on the basis of the information provided by the farmers in the area. Also 13 farmers were selected in Tikuligarh in order to calculate the use of different sources of water for the cropping pattern they practiced for the year starting from 2001 winter crop to 2002 monsoon crop. Pumpage of all the older deep tubewells in the area were obtained from individual water user' group of the deep tubewells as well as from the project. This data included deep tubewell use from 1989. The program CROPWAT was used in order to calculate crop water requirements.

A serious limitation of the fieldwork was that it coincided with a politically very unstable period in the history of the country. Sometimes, I had to postpone field trips due to the tense security situation. The fieldwork was conducted from September 2001 to the August 2002 and then again from February 2004 to May 2004.

In both periods, I travelled from Bhairahawa to the field sites. There were periods in 2001/02 when I was still able to live in the villages. However, in view of the political situation in the country it was not advisable to be visibly linked with any family in the village for the sake of their security as well as for mine. Travelling to the sites was the best option. The good network of roads and transportation facilities made it very easy to travel to the study villages.

Overview of chapters

Chapter Two lays down the context for the study. In this chapter, I present the background of the complex water resources system in Rupandehi district and the study area and describe the modes, methods and processes of intervention in groundwater irrigation as it relates to the Terai in general and Rupandehi in particular. I also examine the social and political institutions and technology that existed in the area before the introduction of groundwater irrigation. I do this because the newly introduced institutions in groundwater irrigation also influence the existing relations between different actors and networks.

In Chapter Three, I provide empirical evidence on the different ways in which farmers have been irrigating in these largely 'groundwater intervention areas'. In order to do so, I examine the historical changes in water use practices in the area, how farmers make use of different sources of water for irrigation, how they define property rights to them and the way they manage the different complexes of water. This chapter shows how technology structured the way farmers had to behave with respect to irrigation but also how farmers themselves, worked out ways for irrigation.

In the fourth chapter I examine the struggles of the farmers in the process of adjusting and incorporating the intervention in groundwater irrigation in three areas inside Tikuligarh VDC. In the first two cases, I examine the reasons behind the different choices that the farmers make even when subjected to similar processes of intervention. In all the cases, the role of the different actors and the strategies they employ, to gain control over groundwater and different sources of water is analysed. This chapter shows that the process of evolution of management around groundwater and its performance can only be understood by looking at how the deep tubewell technology has interacted with: the history of relationships around the different sources of water in the area; the responses to ecological variability; the differences in power structure that existed in the villages before the intervention in groundwater irrigation; the shift in power from one group of farmers to another and the agrarian structure.

Chapter Five is a case study of a village in Madhaulia VDC, and its struggle for securing conjunctive use after the handover of the deep tubewell. In this chapter, I examine the different strategies that the farmers employ in the process of gaining control over both groundwater and surface water and the means they employ to do so. This case study shows how the village is capable of shifting between different institutions for water management at different points in time. This chapter shows that intervention processes are carried out in a dynamic context and that the inserted institutions are susceptible to changes and transformations.

Chapter Six documents water use in Mahuwari where the farmers have opted for only shallow groundwater irrigation instead of surface irrigation in a village outside the deep tubewell project area. The farmers in this area have been irrigating with tubewells, with a minimum role by the state. They devise different strategies to gain control over shallow groundwater, individually and through different social networks. A farmer's choice for irrigation technology is totally driven by their socioeconomic status and they are involved in deriving maximum benefit from agriculture. Most of the farmers are migrants who do not yet have the legal papers that link them with the administrative and political structure in the area and so are not interested to be part of the larger networks that are involved in surface water resources management.

Chapter Seven reviews the major findings of this research. This chapter highlights the significance of understanding irrigation management in an area of complex water resources by looking at the use of different sources of water together. It summarises the findings on multiple use and multiple management of water resources for irrigation and the innovative and active role of the farmers and the power they have in reshaping water source use as they decide what is optimal for them. Implications of the findings for policy in groundwater irrigation and conjunctive use are then presented.

Notes

[1] Deep tubewells in the study area are wells that tap groundwater aquifers from a depth of 150-200m. In Rupandehi district, they are equipped with vertical turbine pumps powered by electric motors that range in capacity from 30kW-75kW.

[2] Shallow tubewells in the plains of Nepal, are wells that go up to a depth of 60 metres.

[3] The Nepal Terai region refers to the southern lowlands of Nepal. They are also a part of the eastern Indo-Gangetic plains. Inside Nepal, all area in the south with an elevation of 30-300 metres is referred to as the Terai. This is the most fertile part of the country and occupies only 17 percent of the total area of the country, which is 147,181 square kilometres. The rest of the country is made up of hills (higher than 300 to 3000 metres) and mountains (all area above 3000 metres).

[4] The country has been divided into 14 zones and 75 districts. Each district is made up of several VDCs. If urban centres exist within a district, they are given the designation of municipalities. There are two municipalities in Rupandehi: Bhairahawa and Butwal.

[5] An ensemble of pump and diesel engine mounted over a shallow tubewell is locally called pumpset.

[6] Irrigation Policy: 1992, 2004. Ninth Plan (1997-2002); Tenth Plan (2002-2007).

[7] Irrigation Policy (2004), in the stresses on the need for promoting conjunctive use of groundwater and surface water irrigation systems, along with different 'non-conventional' irrigation like rainwater harvesting, ponds, sprinklers, drip and treadle pumps. Besides this it also states that 'feasibility for conjunctive use of surface and groundwater shall be taken as the basis for the selection of projects'.

[8] Poudyal and Das Gupta (1987) developed a decomposition model for deriving management policy by integrated use of surface and groundwater in the Tinau river basin; Onta, Das Gupta and Harboe (1991), developed a multistep planning model for conjunctive use of surface-and groundwater resources for Bagmati River basin; Basnet (1993) in her study showed the potential for conjunctive use of ground and surface water in the Sirsia Dudhuara irrigation system in Bara District; Khanal (1994) developed a conjunctive use model for irrigation planning in Narayani zone by minimizing the cost of total water supply, incorporating the stochastic nature of the problem.

[9] Pradhan (1989), Rana (1992) and Yoder (1994) focus on the organization for irrigation by the farmer in the irrigation system. The study by Yoder measures the performance of the FMIS. Zwarteveen and Neupane (1996) focus on issues of gender in irrigation.

[10] Social arenas according to Long (2001: 242) are spaces where different contests and struggles take place across and inside domains. The contests

are over ' issues, claims, resources, values, meanings and representations'

[11] Shah and Singh (2001) state that net area of 13185 hectares were developed under the project.

[12] This database was developed in 1997 in the process of the study conducted by the Winrock International Policy Analysis in Agriculture and Related Resources Program on shallow groundwater irrigation in the Terai. I was also part of the study team.

2

Irrigation and Groundwater

In this chapter, I describe the modes, processes and methods by means of which groundwater irrigation was introduced and how it is used in the Terai and in Rupandehi in particular. The chapter starts with an introduction of water resources in Rupandehi. This is followed by a discussion on the place groundwater irrigation takes in the history of irrigation development in the country and the issues and debates around it. I then focus on the process through which the BLGWIP carried out intervention in deep groundwater irrigation in Rupandehi and on the ways different agencies have been involved in shallow groundwater irrigation. The newly introduced institutions in groundwater irrigation influence the existing institutions and technology in the study area. In the next section, the focus is on how the advent of groundwater tubewells came to take place in a dynamic context where the farmers had already developed ways of managing surface water sources for irrigation. This section, therefore, discusses the evolution of the organisation for managing the surface sources for irrigation and their relation with the study area. The relevant dimensions of agrarian change, the social, cultural, legal and political structures and the influence of these in bringing about changes are also discussed.

Water Resources in Rupandehi

Rupandehi District spreads from the foothills of the lower Churia hill range into the Indo-Gangetic plains. The foothills mark the end of the great mountain ranges to the north and gently slope southward to meet the flat plains of the Indo-Gangetic basin.

Coarse fragmented rocks predominate in the sections towards the northern edge of the Terai adjacent to the Churia Hills to form what is known as the Bhabar zone. This zone is well developed to the south of Butwal by the outwash fans of the Tinau-Dano river system (Gyawali and Dixit 2000). The Bhabar zone sediments extend right up to the northern part of the study area. The two VDCs of Tikuligarh and Madhaulia lie here. However, as one moves towards the south, the proportion of fine material in the alluvial sequence increases and there is a transition to the Gangetic zone of sediments in which beds or lenticles of coarse, clay-free, fragmented rock alternate with beds of lenticles of silt or silty clay. The third site, the village of Mahuwari lies in this zone. The transition from the foothills to this area is short. It is just a stretch of twenty kilometres. That is also the extent of the Terai in this part of the district. The mean elevation of the district is 150 m above sea level, and the topography is relatively flat with gentle uniform north-south gradient. The drainage pattern is dominantly north-south and there is micro-relief between drainage courses (BLGWIP 1999). The temperature in the hot season sometimes goes up to 40°C while the winter temperature falls to below 10°C. The average annual rainfall is about 1600 mm and most of the precipitation occurs in the monsoon (July-September). The other periods are dry with occasional rainfall.

The Tinau is the major river that flows through the district. It drains a catchment in the Mahabharat hills as well as the Churia ranges[1]. After flowing through Nepal, the Tinau joins the West Rapti River in Gorakhpur in Uttar Pradesh in India. The discharge of rivers in these catchments reflects both rainfall and sustained groundwater and subsurface inflow (Gyawali and Dixit 2000). It is different from those rivers that originate either from the Himalayas or the smaller flash flood-prone streams that originate from the Churia hills. Rivers like the Tinau therefore have a more stable flow during the dry season. The run-off from the Tinau is influenced by the rainfall patterns in the upper catchments and unassessed diversion for irrigation by upstream users. According to Gyawali and Dixit (2000) such use is widespread and intense. Floods are also common during the monsoon months. Dry season flow, wherever present, is from groundwater and baseflow contributions. Records of the river flow of the Tinau, even though discontinuous,

FIG 2.1 Average Monthly Temperatures (1971-2002)

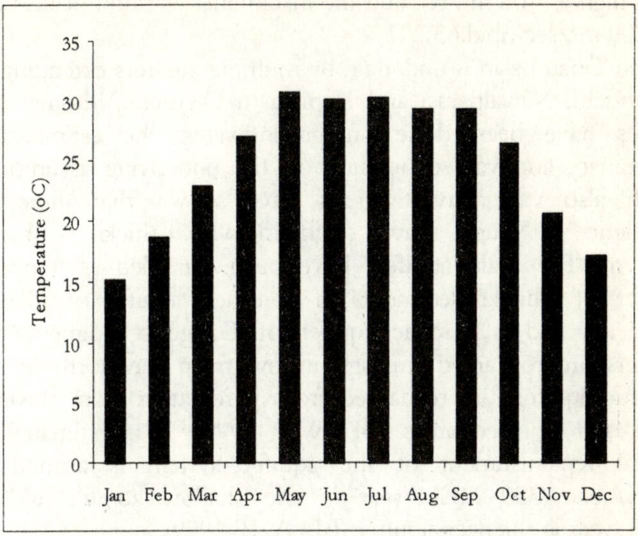

FIG 2.2 Average Monthly Rainfall (1971-2002)

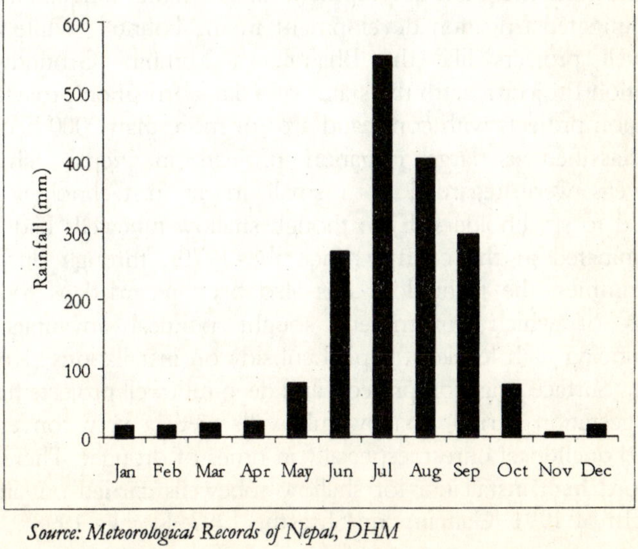

Source: Meteorological Records of Nepal, DHM

show that the mean annual flow is 24 m³/sec. Minimum flow in April is close to 1 m³/sec, the average monsoon flow in August can be as high as 108 m³/sec and the instantaneous flood peak is close to 2200 m³/sec (ibid:63).

The Tinau basin is underlain by multiple aquifers extending into Rupandehi, Nawalparasi and Kapilvastu District. Although many studies have been done on groundwater, the estimates are preliminary and vary depending on the underlying assumptions, which also vary. Investigations have shown that there is a prevalence of 'young' alluvial sediments with a thickness of nearly 1000 m. The main aquifers have been classified as a phreatic aquifer of Bhabar sediments, a confined aquifer of Gangetic sediments and a phreatic aquifer of Gangetic sediments. The aquifers are recharged through inflow from the north and the phreatic aquifers are recharged from percolation from traversing rivers and irrigated areas (BLGWIP 1999). It is estimated that natural replenishment of the aquifer system is around 470 MCM/year (350 MCM/year in the shallow aquifer and 120 MCM/year in the deep aquifer (BLGWIP 1999).

Groundwater Irrigation in the Terai

Groundwater irrigation development takes a rather unique place in state initiated irrigation development in the country. While deep tubewell projects like the Bhairahawa Lumbini Groundwater Irrigation Project rose to the status of a 'large irrigation project' (all irrigation projects with command area of more than 2000 hectares are classified as large irrigation projects in Nepal), shallow tubewells were regarded as a small irrigation technology that catered to smallholders. Even though shallow tubewells had been disseminated in the country since the 1970s, through different programmes, the technology has also been regarded as tool by means of which government sought political advantage by announcing high levels of capital subsidy on installations (Koirala 1998). Surface irrigation projects and deep tubewell projects have a long gestation period. Shallow tubewells were a very convenient tool to quell local unrest, especially in times of drought. There was a record high installation of shallow tubewells carried out in the drought of 1991 (Gautam and Shrestha 1997; Koirala 1998). Most surface irrigation systems are of the run-off-the-river type schemes.

Changes in patterns of rainfall, especially in the crucial paddy planting months have been found to affect shallow tubewell installation by the farmers. Studies have found that the peak season when farmers applied for a shallow tubewell loan was just before the paddy season (Koirala 1998).

There has however been a change in the way how shallow groundwater irrigation was perceived. It has come out of the shadows of the larger irrigation technologies and has become one of the major irrigation development tools since 1997. The Agriculture Perspective Plan (APP) of 1995 focused on groundwater irrigation as one of the major inputs in agriculture that would help revive the failing state of food security in the country. It emphasized that groundwater would significantly help to bring about changes in production in a short period of time that surface irrigation systems had not been capable of. The Ninth Plan (1997-2002) adopted the irrigation development strategies as laid out in the APP.

State investment in groundwater irrigation amounted to 12.2 percent of the total investments made in the irrigation sector from 1956 to the year 2000. The total investment made in that period in irrigation was US$1226 million (Poudel 2000)[2]. Historically, the Department of Irrigation has been involved in the construction of large irrigation facilities. This included both surface irrigation projects as well as the installation of deep groundwater tubewells. The Agricultural Development Bank (ADBN) was the major actor in shallow groundwater irrigation. Then there was a turn of events, after which the Bank was no longer involved in the technical component of groundwater irrigation. The subsidy component of the shallow tubewells was removed and eventually terminated in 1999/2000, under the pressure of the Asian Development Bank during the approval of the Second Agricultural Development Program Loan (Koirala 2001).

A series of reforms were introduced in the irrigation sector after the democracy of 1990[3]. By 1992, the newly elected government initiated a neo-liberal policy, curtailing the role of the state and promoting private sector involvement. These changes were due to the pressure of donors like the World Bank and the Asian Development Bank, pressing for a more market-oriented economics with less government involvement. The 1992 Irrigation Policy reflected all donor concerns and called for more farmer participation in irrigation management. It called for the transfer of state-constructed irrigation systems to water user groups. Reforms in agriculture since 1990 include the removal of subsidy on fertilisers and the de-regulation of fertilizer trade and distribution. The private sector has been allowed to participate in distribution of the fertilizers. Removal of subsidy on shallow tubewells was part of this move too. The responsibility for shallow tubewell irrigation went to the DOI, which previously was not involved in these smaller irrigation technologies.

After the process of removal of subsidy on shallow tubewells, the Asian Development Bank (AsDB) financed the Community Groundwater Irrigation Sector Project (CGISP). The project is being implemented with a target of providing irrigation facilities to 110000 small farmer families in eastern and central Terai districts. It started from the fiscal year 1997/98 under an Asian Development Bank loan. The aim of the project was to cover 300 VDCs and provide irrigation to 60000 hectares. It aimed to install 90 percent of the shallow tubewells on a group basis and the rest as individual tubewells. In addition, the project included construction and upgrading of village access roads in all the VDCs. Non-governmental organisations (NGOs) were involved in the project for farmer mobilization and project implementation was done by the consultant. This is the most recent large scale investment in shallow groundwater irrigation amounting to 42 million US$. The Nepal Irrigation Sector Project financed through the World Bank loan is implemented in groundwater irrigation in the western Terai. Groundwater installation in Rupandehi district by the DOI is carried out by a field office in the district. The groundwater field offices (Groundwater division of DOI and GWRDB) were established to provide technical assistance and services to ground water development projects.

As the scope for the construction of large-scale projects in the Terai has diminished, the trend in irrigation investments is geared towards groundwater irrigation and rehabilitation of existing government-constructed projects and FMISs. In addition, projects and programmes implemented by the donor agencies through non-governmental organisations have also started installing shallow tubewells in the Terai.

There is a wide range of difference in the cost of groundwater technology. This depends on the size of the technology. The cost per hectare of the BLGWIP deep tubewells was US$ 5174. Shallow tubewells cost an average of US$ 300 while the cost of treadle pumps was US$ 230 per hectare (Shah and Singh 2001). Deep tubewell projects, like all surface irrigation projects, have been provided with capital cost subsidies by the government. The basis for this priority is the high initial costs associated with it. The removal of subsidy from shallow tubewells was criticised by the proponents for subsidy re-instalment. They argue that the government has been biased because it has been subsidising larger irrigation projects while neglecting the issue in case of poor-friendly smaller irrigation technologies[4].

Groundwater irrigation in Nepal, as in other water-abundant areas of the Indo-Gangetic plains, is being carried out with the view of increasing the access to water by the farmers. However, emerging issues from greater experience in groundwater irrigation in the Terai as has been stated in Chapter one, is that there is a very low level of use of the installed groundwater structures. This low level of utilization has been attributed to two factors: a low rate of expansion of groundwater services and optimum utilisation of the installed facilities. The first has been related to the low demand of tubewells by the farmers and also to the inability of the government to increase the number of installations. The low level of use of groundwater structures has been attributed to decreasing land holding size, the inability to go in for crop diversification and the indivisibility of groundwater tubewells. The tenth plan (2003-2007) reviewed the status of groundwater irrigation in the period of the ninth plan (1997-2002). It was the ninth plan that first made groundwater irrigation into a development priority. The tenth plan states that there has been a massive decline in the demand for tubewells by the farmers since 1997. The plan cites several reasons for this trend: 'sharp fall in the price of rice and wheat in the national and international market, unrestricted flow of cheap Indian food grains through the open border, phase-wise withdrawal on the subsidy on shallow tubewells, and the rise in fuel prices'.

The other issue arising naturally with groundwater irrigation is that of fuel prices. The major sources of energy in use in groundwater in the Terai are electricity and diesel. All deep tubewells of BLGWIP are run by electricity while the shallow tubewells use diesel engines. The rate for demand charge or the flat rate of electricity has more than doubled since 1996. In 1999, the rate increased from NRs. 20 per kVA to NRs. 26. It again rose to NRs. 47 per kVA in 2001. The power tariff rose from NRs. 3.05 for unit consumption in 1996 to be NRs. 3.25 per unit consumption in the year 1999. It rose again to NRs. 3.40 in 2001. The cost of diesel rose from NRs. 11.90 per litre from 1994 to 13.40 in 1996. In 1997, the price was NRs. 14.40 per litre, NRs. 28 in 2001 and NRs 36.40 in 2003. The interplay of energy pricing and the choice between different sources of water, and how it shapes irrigation water management and resulting institutions is also a topic of interest in this study.

Deep Groundwater Irrigation

State involvement in groundwater irrigation started in 1969 with the initiation of the project 'Groundwater Resources Investigation in Nepal Terai' (1969-74). This was carried out under the grant of USAID. It was in the same years that several donors started the exploration of the potential of water resources in the country[5]. Test wells and tubewells were drilled for the collection of water samples for analysis. After the end of this project, a feasibility study[6] was carried out. This led to the start of the Bhairahawa Lumbini Groundwater Irrigation Project.

The history of deep tubewell irrigation in Nepal is synonymous with the Bhairahawa Lumbini Groundwater Irrigation Project (BLGWIP). Implementation of this project started in 1976 under a WB/IDA loan. By this time several bi-lateral and multilateral agencies had already implemented deep tubewell projects of this nature elsewhere in other parts of the Indo-Gangetic plains[7]. Nepal had opened to the outside world only after 1951. The construction of large-scale irrigation projects started only in the sixties and the seventies in the country.

BLGWIP takes quite a unique place in the history of agency-initiated irrigation development in the country. It is the biggest groundwater irrigation project as well as the most expensive irrigation project implemented in the country so far[8]. The cost per hectare for the deep tubewells of the BLGWIP amounted to US$ 5174. Besides being the biggest groundwater irrigation project[9] in the country, it also falls in the category of large irrigation projects along with other surface irrigation schemes. It can therefore be said that the deep tubewells have always received similar priority as large scale surface irrigation projects in Nepal.

The project installed 182 deep tubewells in Rupandehi district in twenty-five years period. It was implemented in three phases. It installed 65 deep tubewells in the first phase. These tubewells became operational in 1982 to 1988. Thirty-eight deep tubewells were installed in the second phase of the project. The final phase of the project was implemented in 1990 and went up to 1999. A net area of 13185 hectares was developed[10] for irrigation. The project also included a 92 km long rural road and electricity transmission. All these have substantially transformed the rural landscape in the district. It has also encouraged more migrants to come and settle in the area.

Parts of Tikuligarh and Madhaulia VDC fell under the project target area in the first phase period of implementation. These VDCs therefore have some of the oldest tubewells in the area. Several villages that irrigated from the Sorha and Chattis Mauja surface kulos were also designated to be part of the BLGWIP. Several villages in Tikuligarh and Anandban VDC that were irrigating from Sorha Mauja fell on the project area. Similarly, a part of Madhaulia VDC that irrigated from Chattis Mauja and parts of Gangolia and Karaiya VDC, also became part of the BLGWIP. Parts of Tikuligarh VDC latter had deep tubewells installed during the third phase of the project. None of the villages in this study fall under the second phase of the project.

BLGWIP was implemented in twenty-five years. The project started with intensive state involvement that was typical of the 1970s trend in irrigation development worldwide. Until 1989, design was done without any farmer participation (Olin 1994). Later on, the project incorporated a more 'participatory approach' that was characteristic of irrigation reform policies in the 1980s and the 1990s. This included the incorporation of the 'demand based participatory approach' in project design for the deep tubewells installed in the third phase of implementation. All deep tubewells that had been installed before 1992 were to be transferred to the water users' group. The project worked inside a contiguous project area for the initial phases of implementation, more or less determining the location of the tubewells. In the third phase of implementation, it did not work within a contiguous project area, but instead opted for a 'demand-based participatory approach'. The 'participatory concept' for BLGWIP was developed in 1989

(BLGWIP 1999). In this stage, the farmers who had not obtained a deep tubewell could request the project for a tubewell. Some parts of Tikuligarh VDC were not designated for installation in either the first or second stage of the project. Four more deep tubewells were installed within the boundaries of this VDC in the third stage of this project.

There have been changes in the design of the deep tubewells as well. The older tubewells in the project area are equipped with high capacity motors of 75kW and an open flow conveyance system. In the third phase, the deep tubewells are equipped with lower capacity motors of 35kW. The distribution system includes underground pipe distribution system. In all cases each unit of the deep tubewell installed was designed to operate as a separate single irrigation unit. The average design command area was around 120 hectares.

All investment costs were borne by the project. In addition to this, all operating, maintenance and repair costs were borne by the project authority until 1992. The pumps were operated by a pump operator hired by the project. The farmers of Stage I deep tubewells had four years of unlimited free water. By this time, the first set of tubewells under Stage I of the project had already been installed. Farmers had already started using them at no cost. Some remaining tubewells planned for Stage II of the project were still to be installed. According to Tahal (1989), the consulting firm engaged with the project throughout its twenty-five history, the farmer participation concept developed at that time envisaged two processes: one for the areas already under irrigation and another for new areas. A Farmer Organisation Division was established in the project in order to implement farmer participation in some remaining tubewells in the second stage and for the incoming tubewells in the third stage. This was also to prepare the stage for the transfer of the deep tubewells to the farmer groups. The project deliberately termed this as 'takeover' to inject a 'feeling of ownership' for the farmers (BLGWIP 1999b) such that the farmers felt that they were taking over something that belonged to them. However, the Nepali term that has been in use by the farmers is *hastantaran*, which means 'handing over', showing the difference in conceptualization of the process by the two sides. The process of transfer of tubewells met with different reactions in different tubewell areas. There has been no systematic published documentation on the experience of individual tubewells[11].

Project documents and other reports by those involved in the project state that the process of transfer was not a smooth one. Olin (1994) describes the processes and experiences of the transfer programme in the first and the second stage tubewell areas. The takeover concept was basically for 64 deep tubewells of the first stage and 16 completed tubewells from the second stage. In a survey conducted in 1992 by Tahal, ninety-two percent of the farmers in stage one of the project stated that they preferred the government continue to operate the deep tubewells.

The project established a division chief under the Farmers Organisation Division (FOD), with association organizers (AO) and AO supervisors. The responsibility of the staff in this division was basically to prepare the farmers for 'participatory concepts' and train the water user groups on how to handle the deep tubewells upon transfer. A 'Guide to Farmer Participation' was prepared. The AOs were trained to create awareness about participation in existing tubewells, as well as to assist in organisation of the new WUGs.

The time of the transfer process coincided with the time when the country was undergoing a political upheaval: the democratic movement of 1989/90. Both the first and the second stages of the project had been implemented during the Panchayat[12] period. The handover process started at that period in the history of the country, when the people were free to express their thoughts, actions and political affiliations. At first, the farmers were not willing to take the responsibility of the tubewells. The process therefore met with great resistance from the farmers. Petitions were made by the farmers to the government asking for the withdrawal of the transfer process. The farmers threatened the government and the project that all chairmen of 38 WUGs would resign, claiming that they were not capable of managing the deep tubewells.

There were a series of negotiations and confrontations. The project then switched over from a voluntary mode to the use of pressure. In the summer of 1992, the project decided to close the deep tubewells just before the paddy season in order to make the farmers nervous. Project authorities considered that this action would make the farmers ready to take over the deep tubewells. However it did not work because of intense pressure by the farmers. The project finally had to open the tubewells. The deep tubewells were open till the beginning of November that year.

The project took up another tactic just before the wheat growing season in the winter of the same year. The staff of the project responsible for the handover process locked the deep tubewells again. After doing this, the staff then put forth the conditions of the project. They told the farmers that the project would line the canals of the BLGWIP if the farmer groups decided to take over the deep tubewells.

The farmers had not paid for demand charge since the time the project had locked up the tubewells for the first time in summer. BLGWIP had already started setting the payment of demand charge as part of the process of 'institution building' before handover. The farmers informed BLGWIP that they would pay the demand charge for five months but not for the month that the project shut down or closed the deep tubewell. They also agreed to pay fines for the same months. By this time the project had already induced one tubewell in another VDC to overthrow its deep tubewell chairman who was resisting the taking over of the well. The new committee had then elected a new chairman, and taken over the tubewell. After one tubewell had been taken over by the farmers, the project took this as symbolic for the whole project. They claimed that the handing over was complete. All deep tubewell committees had to take over the keys of the pumphouse. The deep tubewell water users' groups were not immediately handed over the keys to the panel board of the electric component by the project. Organized theft occurred in many tubewells around the district, involving the removal of the auto-transformer. They were found sold across the border in India.

The farmers declared that they had been bluffed. Project documents cite there was 'resistance to take over the deep tubewells by the farmers of the Stage I area, who accused the project of political bias in forcing handing over of deep tubewells' to the other farmers while they were still paying nominal water charges (BLGWIP 1999a: 46). The farmers in the stage I area had to pay the water tax[13]. The project had induced the handover in a deep tubewell that was installed in the second stage of the project.

The farmers, who (in their own words) were used to opening the pump for a small activity like bathing, had to prepare themselves to sustain the expensive deep tubewells all on their own. Taking over the deep tubewells involved many things for the farmers. This included taking up the responsibility for all costs for repair, maintenance as well as of operation. They were first of all particularly opposed to paying the demand charge (flat rate) of electricity to Nepal Electricity Authority (NEA). This is the tariff which they had to pay whether they used the tubewell or not. The farmers were willing to pay for the operating charge. Operating costs were something that they 'believed' in. According to the farmers, 'we pay for something we use, not for something we do not use'. This became a political issue. The tariff for agriculture was then revised in March 1993[14].

In addition to this, the farmers were also apprehensive about the high costs that they would have to pay if they were to replace the transformer coils that cost more than NRs. 200,000[15]. Electricity had been brought into the area specifically for the deep tubewell project by the NEA's 33kV lines to four substations where it was stepped down to 11kV and distributed to all deep tubewells. When the project was in control, the electricity charges were paid by the project authorities. That means the project was basically a client of the NEA. After transfer of the deep tubewells to the farmers, the NEA still kept its stand and rigidity in viewing the deep tubewell project as its private client. However, after the transfer process, each and every deep tubewell was responsible for its own operation, maintenance and repair. The NEA then regarded them as individual clients. The farmers demanded that the transformers be taken over by the NEA so that they would not have to be responsible for the costs incurred around it. Revenue obtained from all the deep tubewells in Rupandehi fetches a lucrative amount of money for the NEA. It would fetch NEA slightly less than US$ 60000 per year from the 182 deep tubewells when only the amount based on collection of demand charge is calculated.

In his paper on the experience of processes of management transfer of the deep tubewells, Olin (1994) writes that the 'whole process was characterised by conflict and confusion'. He lists certain events that have led to this situation. First, there was the confusion caused by policy changes in the course of implementation. This, according to him, was one of the reasons for resentment among the farmers. The government had implied that the water tax and takeover program could work as alternatives. But according to the project, this was not the case. The next cause of the confusion between the farmers and the project was on the offer for canal lining. The project was not able to line the canals completely as promised. There was a change in irrigation policy (1992) between the time of the promise made by the project and the time the project could implement lining the canals. The irrigation policy incorporated 'farmers' participation' whereby the farmers were required to contribute a certain amount for rehabilitation works. The farmers were not willing to comply with this rule. Then confusion again arose when there was a budgetary cut of the government. This decreased the budget allotted for canal lining as well as the budget that had been set for the AOs. The project could not line the whole length of the canals as had been promised earlier (Olin 1994). The farmers in the area still complain that the project cheated them because they did not line the canal in the length promised to before handover. From the project point of view, these uncertainties hampered their process of transfer of deep tubewells.

Each deep tubewell has a water users' group. This group consists of all the water users within a deep tubewell. This body elects an executive committee. This committee comprise of a chairperson, vice chairperson, a secretary, a joint secretary, treasurer and two to six additional members. The chairperson could be elected by a secret ballot, while others were elected by open ballots. The general assembly sets up rules for O&M, fixed and variable fees and membership fees. It also elects the executive committee and approves annual programs and budgets. The executive committee implements the rules and programs, prepares income and expenditure statements and performs all administrative duties. It nominates new members, proposes fixed and variable fees to the general assembly, pays bills and oversees water management and distribution.

The chairpersons are in charge of the overall activities of the deep tubewells. They have the right to deny water to those farmers who do not pay their fees. If there is any type of construction work involved, they are responsible for mobilising the farmers for cash and labour. They also coordinate the work of the pump operator and settle conflicts related to water. Pump operators were hired by the project when it was in control. Presently a local person is chosen by the executive committee of each individual deep tubewell.

The project set up Regional Coordination Committees in order to assist the farmers in different WUGs in procuring maintenance equipment and services, in four zones of the project area. There is also a Central Coordination Committee set up by the project in 1998 to coordinate the activities of the regional committees and render support in common procurement and O&M problems to regional committees (BLGWIP 1999). It consists of five executive members and six members.

A deep tubewell user has to pay two types of charges to the WUG. One is the fixed charge and the other is a variable charge. The fixed charge is the demand charge that the tubewell has to pay to the NEA every month. This depends on the capacity of the electric motor. Each farmer has to pay this in proportion to the land that he or she owns within the deep tubewell area. The variable charge is the amount that each farmer has to pay depending on the time of irrigation.

It has been mentioned above that the project changed its approach in the later stage of implementation and adopted a 'demand driven' approach. In the latter tubewells, the project demanded that if any village or group of farmers 'wanted' a deep tubewell, it had to apply for it. The project sent AOs to 'motivate' the farmers for irrigation and to explain the pros and cons of deep tubewell irrigation. According to the project reports they 'sold' the idea of irrigation to the farmers and helped them organize before the system was designed or built. The farmers in the villages that did not have a deep tubewell in the first and second stages of the project had been observing the whole process of project implementation. They were already aware of the costs that groundwater irrigation would entail. When the project initiated the third phase, one of the things that they could sell to the farmers was the changes they promised in the design: smaller capacity tubewells and a different distribution system. This would be cheaper to run than the older tubewells. Under this approach the farmers had to submit an official request for irrigation after they organized themselves in a water user's group. There were set agreements defining rights and obligations of both the parties: the farmers and the project. The farmers were supposed to be included in all stages of the design and participate in the construction of the tubewell. The government contributed the largest part of the costs while the farmers had to contribute a nominal share.

Shallow Groundwater Irrigation

The main actor in shallow tubewell irrigation in the Terai up to the late 1990s had been the Agricultural Development Bank Nepal (ADBN). The total number of shallow tubewells that had been installed by the bank in Rupandehi was 3777 (Well Certification Records, ADBN). When this number is calculated with respect to the area that is technically suitable for shallow tubewells (which is 810 sq. km[16]) the tubewell density comes at 5 per sq. km for this district. This figure is second only to Bara District in central Terai. This figure is based only on the number of wells installed by the Bank. The process of implementation of shallow tubewells by the bank was the same all over the Terai. So, unlike the process for deep tubewells, I do not describe the process in particular reference to Rupandehi. I do so with reference to the overall approach taken by the bank. Besides the bank, other organisations were also involved in groundwater irrigation. The Farm Irrigation and Water Utilization Division (FIWUD) under the Ministry of Agriculture as well as the Ministry of Panchayat and Local Development were also involved in groundwater irrigation. Lately the non-government sector has also emerged on the scene.

ADBN promoted shallow tubewell irrigation through loan assistance and technical support. This process started from the seventies until 1999, after which the Bank moved away from the technical component of shallow tubewell irrigation. The level of subsidy for the tubewell and the pump/engine or pumpset was 40 percent of the cost for individually owned shallow tubewells and 75 percent for group-owned shallow tubewells until 1997. Later it was reduced to 30 percent for individually owned shallow tubewells and 60 percent for group shallow tubewells. The subsidy was completely removed in 1999.

One of the criticisms of the implementation process of shallow tubewells by the Bank has been its focus on the number of shallow tubewells that could be installed in each Terai district. Each year, the Bank was given the target of developing irrigation facilities by the government. The implementation of larger irrigation schemes went to the DOI, while small-scale irrigation development was also the responsibility of the Bank along with some other agencies like the Department of Agriculture and Ministry of Local Development. The subsidy quota that the Bank allocated for groundwater irrigation was then distributed among its various field offices.

In 1985, a study[17] was conducted, which calculated the potential number of shallow tubewells that could be installed in each of the twenty districts of the Terai. This study had been the basis by means of which the ADBN carried out its programme on shallow tubewell development. According to this study, the number of shallow tubewells that could be installed in Rupandehi District was 3807. The Bank tried to install the stated number of shallow tubewells in the district. Shallow tubewells were installed in whichever site the farmers requested, as long as they found an aquifer. In the absence of strict groundwater legislation, it did not matter where the shallow tubewells were installed. Control was exercised through the ADBN, which required a minimum spacing of 200 meters between wells. However in practice this was also breached. There are no regulations on private installations. There was an understanding between the BLGWIP and the ADBN that the latter would not 'trespass' on the boundaries of the former. But installation of shallow tubewells totally depended on the relationship between the farmer, driller and the field staff. Groundwater is assumed to be under private ownership.

Various criteria were set up by the bank through which farmers could qualify for a shallow tubewell. Farmers could obtain a shallow tubewell individually or as a group. Programmes such as the Small Farmer Development Programme also focused on group tubewells. The subsidy was available for drilling the tubewell and the pumpset. A farmer who wanted to apply for a tubewell loan was eligible to apply for one, provided he or she had a minimum amount of contiguous irrigable land of 1 hectare. This land was kept as collateral in order to obtain a loan. This criterion for 1 hectare of land was followed for many years. It was fixed for a tubewell of four inches diameter. This requirement was gradually relaxed for smaller diameter tubewells. This contiguity of land holding was found to be interpreted and applied by different field offices in different forms (Koirala 1998). So sometimes a farmer who had less than 1 hectare obtained a shallow tubewell, while in other cases they did not. The minimum land requirement for 'community tubewells' was set at 4 hectares. At least three farmers who owned 4 hectare of contiguous land had to come together to apply for this loan. This requirement is very stringent when compared to individual applications. In order for three people to have a contiguous 4 hectare of land, they will need to have an average of 1.33 hectares each. This is 33 percent more than for private ownership. This is also the reason why, despite a high level of subsidy, the number of 'community tubewells' installed by the bank has been less than 4 percent of all the tubewells installed by the bank across the Terai). Even those schemes which were obtained through the community scheme ultimately turned into individual ownership (Gautam and Shrestha 1997; Koirala 1998).

The other programme that has recently been involved in installation of shallow tubewells in the study area has been through the Implementing Priority Productive Investments phase of the Participatory District Development Programme of the UNDP. In this programme, community organisations (COs) initiated by the programme invest in small-scale productive infrastructure development such as irrigation projects, drinking water supply schemes, micro-hydro, development of higher levels of cooperative enterprises, commercial forestry etc. The formation of COs is initiated through a sensitization programme, or a series of "dialogues," that forges partnership with the local communities. The CO adopts a community-oriented constitution to govern the management of the CO. The constitution spells out the management structure of the CO, which includes a Chairperson, a Manager/Secretary, and the rest as members. The CO is then registered at the VDC in order to establish a formal linkage. Each CO chooses its Chairperson and a Manager for the smooth execution of the organisation activities. This is done through the concurrence of all CO members. The COs meet every week to discuss development issues of mutual concern to all the community-members. During their meetings the group members deposit their monetary contributions. Every member saves an equal amount each week to allow for cooperative decisions among the members. The capital collected in this manner is the collective asset of the CO. This fund is used for loans to finance any micro-enterprise development at the household or village levels. Thus COs can lend money to their members at a rate of profit which satisfies local credit needs, and at the same time supplements the CO savings, and women too have control over their savings. The savings of the COs is only a form of collective asset, and more like a membership fee that the members have to pay to reap the benefit of being part of a collective association. The organisation, therefore, is different from that of a savings and credit organisation.

Groundwater Tubewells in the Study Areas

The deep tubewells in the study area are wells that go up to a depth of 150 to 200 m. They are equipped with vertical turbine pumps powered by electric motors. They range in size from 30 kW to 75 kW, designed to irrigate a similar range of area of around 120 hectares (Table 2.1, 2.2). The design of the deep tubewells that were drilled at the beginning of the project is different from those that were installed in the final stages of project implementation. The older deep tubewells are equipped with larger capacity motors and have an open flow distribution system. The new tubewells have a low-pressure underground pipeline distribution system conveying and distributing pumped (or artesian) flow to outlets each serving an average of 3 ha each. The channels of the older wells are fully lined in some deep tubewells and partially lined in others. Each stage or phase of the project had its own design characteristics associated with it. The deep tubewells in the study sites fall in the first and third phase of the project.

The average discharge of the bigger tubewells is 0.111 m^3/sec. These deep tubewells were designed with a view that all farmers would need irrigation water at around the same time and that it would be distributed along each distribution channel. In the new deep tubewells, water is lifted to an overhead tank. From here it gets divided into four loops and flows through the loops. Each loop has ten gated outlets. These outlets are controlled by a riser valve. Each outlet serves around 3 hectare each. The farmers organize at the outlet and distribute water through the ditches they have constructed. The pump house also includes living quarters for the pump operator. Tubewells number 3, 5, 6 and 9 in Table 2.1, are the tubewells discussed in chapter four. Tubewell number 4 in Table 2.2 is the tubewell in chapter five.

Shallow tubewells in the Nepal Terai, are wells installed up to a depth of 60 m. Local methods are used in to drill these wells. These local methods of drilling are locally called 'boring'. These were recognized by the ADBN and incorporated into its implementation

TABLE 2.1: Characteristics of the deep tubewells in Tikuligarh VDC

Tubewell	*Head (m)*	*Motor Rating (kw)*	*Discharge (m3/hr)*	*No. of households intended to be served initially*	*Year Installed*	*Design command area (hectares)*
1	15.0	75	520	150	1982	138
2	19.0	75	460	140	1982	123
3	17.0	75	435	155	1982	117
4	19.0	75	470	114	1982	138
5	12.0	75	415	114	1982	110
6	20.0	75	545	123	1982	150
7	20.0	75	400	140	1982	127
8	24.0	75	500	125	1982	133
9	26.0	37	300	148	1999	132
10	27.5	37	300	157	1999	124
11	25.0	37	300	120	1999	140
12	22.5	30	300	144	1999	138

Source: BLGWTP, 1999

TABLE 2.2: Characteristics of the deep tubewells in Madhaulia VDC

Tubewell	*Head (m)*	*Motor Rating (kw)*	*Discharge m 3/ hr*	*No. of households intended to be served initially*	*Year Installed*	*Design Command Area (hectares)*
1	24.5	75	400	120	1982	106
2	10.5	75	455	155	1982	121
3	15.2	75	440	150	1984	118
4	25.0	75	600	155	1985	145
5			Defunct			106

Source: BLGWIP, 1999

programme. The Bank had adopted five methods of drilling[18] across the Terai. The choice of the method depends on the subsurface conditions in the area. Local drilling mechanics identify the aquifers and drill the wells in the farmers' fields. According to the drilling mechanics in this area, they apply the Thokuwa method to drill up to 30 feet depth. To drill up to 300 ft to tap the flowing artesian, the drilling mechanics use the Bogi. The cost of drilling a shallow well depends on the depth to the aquifer. It ranges from NRs. 6000 to 10000. The cost for free flowing artesian wells is around NRs. 14000. These free flowing wells mainly supply drinking water.

The design of the tubewells across all the study area is uniform. Each shallow tubewell unit has a tubewell of four inches diameter, mounted with a diesel engine and pump (collectively referred to as a pumpset). The capacity of the diesel engine ranges from 5 to 7 hp engines. All shallow tubewells in the study area were equipped with a hand pump for priming purposes. Water is pumped and carried to the field by means of earthern channels. Irrespective of whether a shallow tubewell is used individually or by a group of farmers, the design of the shallow tubewell is the same. The diesel engine of the shallow tubewell unit is also used for threshing purposes. It is mounted with a thresher when used for this purpose. There are also free flowing artesian wells in Madhaulia and Mahuwari. The diameter of the pipes installed to tap the free flowing artesian[19] well is of 1.5 to 2 inches.

Groundwater Use in the Study Area

It has not been possible to present data on the extent of use of shallow tubewells in the study area. Shallow tubewells are operated privately by the farmers and a full survey and documentation of their use has not been made. The WUGs of the deep tubewells keep a record of the groundwater use in the area, which was also documented by the project. I present a review of the extent of deep groundwater use by the farmers from 1989 to 2001 in two of the study sites: Tikuligarh and Madhaulia VDCs, based on the information provided by both the farmers and the project (Figure 2.3). The graph shows groundwater use from older sets of deep tubewells that were installed by the project in the 1980s. Four other deep tubewells were installed by the project in Tikuligarh VDC by the end of 1999. The figure does not show the use of these deep tubewells[20].

The graph in Figure 2.3 shows a drastic decrease in deep groundwater use since 1996. From this year, groundwater use from most of the tubewells was at an average of only 0.5 MCM. The figure shows that farmers reduced the use of deep groundwater after the transfer process. Deep groundwater use was much higher before 1992, when the project was still managing the tubewells. The sharp peak in 1991 indicates groundwater use in a drought year. The figure also shows that there was a variance in deep tubewell use even before the transfer process. Farmers in several tubewell areas used relatively less water than others.

The Agrarian Structure and Surface Irrigation

The agrarian economy of the country, from the time of unification in 1768 to 1951, was characterized by a feudal system. The king or those who represented him had the power to control all land and other productive resources within the country (Regmi 1976)[21]. The Ranas took power from the Shah kings in 1846 and ruled the country for 104 years. The main source of revenue for the rulers came from control over land. They assigned tracts to land to different subjects under various tenurial conditions. According to Bughart (1997), royal and state treasury were not considered separate before 1951 (Pradhan 2000:46).

The majority of the kulos that were later called farmer constructed and managed irrigation systems emerged out of the land tenure system prevailing at the time (Benjamin 1994; Pradhan 2000; Khanal 2003). They either emerged out of the *birta* and the *jagir* system of land tenure or were further expanded through these systems (Benjamin 1994; Pradhan 2000; Khanal 2003). According to Pradhan (2000), the organisation of these kulo systems were based on the rules that regulated the relationship between feudal lords and those who cultivated the land. He further states that these very rules that were considered external, later were incorporated as internal rules of the community through the passage of time. Both the birta and the jagir system of land tenure were forms of privileged landownership that assigned land to people favoured by the ruling class. The rulers therefore assigned tracts of tax-free land to those individuals or groups of people that they favoured. This was a form of power that the rulers exercised in order to gain political and administrative authority. It was also a

FIGURE 2.3: Deep Groundwater Use in Study Area (1989-2001)

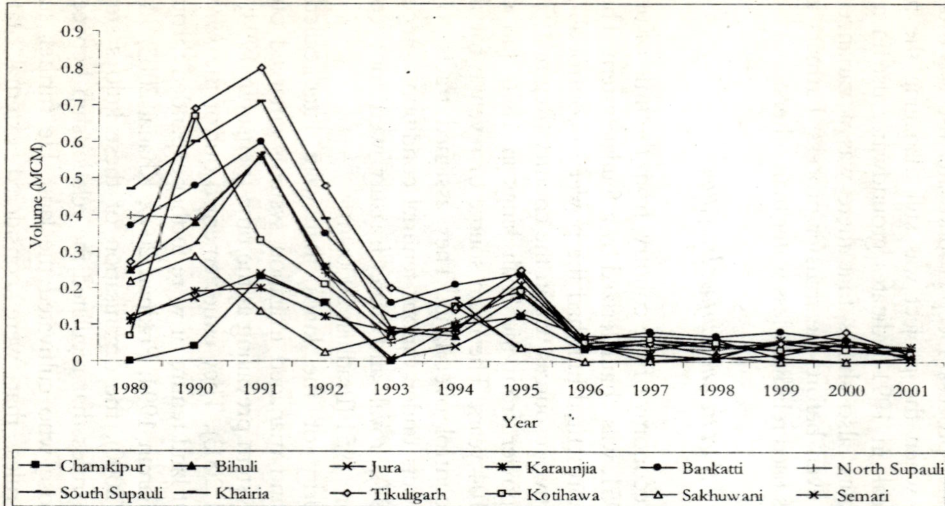

Note: DTW Chamkipur is actually the well referred to in Chapter Five as Bihuli well.

a) Source: BLGWIP, 2002

b) Source: WUG of respective deep tubewells, 2002

tool whereby they could control those disloyal to them. This was because they also had the power to either grant or take away the land grants. This happened when there was a change in leadership. Birta was mostly conferred to priests, religious teachers, soldiers and members of the nobility. It was granted to those members of society who by tradition could not take part in economic pursuits. According to Regmi (1976), they were therefore maintained at the cost of the agrarian population.

Birta grants were also made to persons who organized land reclamation and settlement projects and functioned as revenue collectors on behalf of the government. Irrigation canals were either opened up or extended in those areas that already were being irrigated.

The jagir system of land tenure arose in a directive given in 1793 when the king gave military employees cultivated land as remuneration instead of cash. According to Regmi (1971), the system came up because of the lack of a broad-based money economy and public finance system. There was an increase in area under jagir system in areas that were of strategic importance to national security. This was mostly in view of the possible war with British India (Regmi 1971). According to Regmi (1976), the people who received such grants had to take the responsibility to promote settlements and reclaim wastelands. As one of the objectives was to organise military bases, they were also responsible for food supply; in charge of transportation of arms and ammunition as well as to collect information about developments in India.

Both birta and jagir land grants entitled the beneficiaries to collect revenue from all sources in the area controlled by them. In addition, the people who were given the land grant also had the power to dispense justice and to demand unpaid labour (*jhara*) from the cultivators. They also had rights to a share of the agricultural produce as well as the right to impose levies and taxes on use of forests, pasture and water by the common man. This gave them the power of control not only over the land but also the inhabitants living and cultivating on it.

Another form of tenure also existed at the time in the Terai: jimidari. The king selected the jimidars[22] as revenue collection functionaries. They were assigned land under *jirayat*[23] tenure. They were also given a percentage of the revenue that they collected. However, unlike the *zamindars* of India they were not given ownership rights to the land under their jurisdiction.

The basic unit of land administration was the *parganna*, which comprised a number of villages. A functionary called the Chaudhary was appointed from among the local landowners to collect the revenue. The function of the tax collection was entrusted to selected jimidars at the village level or Chaudharis at the parganna level. The jimidars, therefore, acted as the authority between the local level and the centre. 'The objective was not only to collect land-tax, but also to create a rural aristocracy capable of injecting capital investments and entrepreneurial ability in the field of agriculture' (Regmi 1976:108).

The form of system that existed in the western districts of Rupandehi, Kapilvastu and Nawalparasi was different from the jimidari system of the other Terai areas. The system that existed here was known as ukhada[24]. In the ukhada system, landownership rights were vested in the jimidar. The jimidar would collect rents in cash from registered landholders. The profit he made was the difference between these cash rents and the tax payable to the rulers. Wasteland and forest land was registered in the names of the jimidars on a taxable basis. In order to lessen the burden of tax payment obligations, the jimidars, gave lands to cultivators on relatively favourable terms. This, according to Regmi (1976) was termed as Ukhada. Since rents were paid in cash, the tenant profited from rising prices. This was very different from the systems in the other parts of the country.

According to Pradhan (2000), there were no regulations that defined how the jagir and the birta holders were related to their tenants before the 1854 *Muluki Ain*[25]. The Muluki Ain was the major source of state law between the period 1854 and 1951. Tenants were registered in the tax assessment records to ensure security of tenancy. The Muluki Ain granted more rights to the those who held the birta[26], but the revision of 1906 of the Ain gave more rights on the tenants (Pradhan 2000). Tenancy rights to birta lands could be transferred as long as it did not affect those who held the birta.

Tenants had to provide free labour whenever required. The entire adult male population irrespective of class, ethnic group or caste[27] was under the obligation to render compulsory and unpaid labour services whenever required by the state/king (Regmi 1976; Benjamin 1994; Pradhan 2000). The people who were given the authority for land reclamation were given the right to demand labour from the people in different kinds of activities like the construction of temples, irrigation channels, embankments, erosion and river control works (Regmi 1976). Benjamin (1994) writes that the irrigation systems in Nepal were constructed by tenants of birta and *raikar*[28] lands and probably by slaves. Officials were sent to eastern Terai districts from time to time after 1793 to make arrangements for irrigation facilities in order to promote land reclamation and settlement (Regmi 1971).

Traditionally, the landlords in the study area used to keep labourers known as the '*haruwa*'[29]. When the jimidari system was still prevalent, the landlords sent these labourers for canal cleaning operations, which they had to do it without any pay. This system of keeping permanent labourers does not exist in the same form as it did before[30].

Land reform measures were introduced in the country in the 1960s[31]. Estimates in the 1960s showed that landlords owned two-thirds of the cultivated land in Nepal: three-fourths of the farm population in eastern Terai owned no land at all (Skerry 1964). There were changes in the organisation of irrigation kulo irrigation that relied on compulsory labour for its upkeep before this period. Changes had already started in the way surface irrigation was managed in this period. One of the factors influencing these changes in Rupandehi was the increase in migration into the district.

Formally, the imposition of a ceiling on land ownership broke down the existing concentration of landholding at least in theory. However, the land-reform program had little impact on the agrarian structure as such.

The campaign of identifying tenants was spread over three successive years. In those areas where the campaign had yet to start, the landowners had ample time to evict the tenants and claim that they were the tillers themselves. Many tenants were also left out in the process. Later attempts by the Land Reform office to give land ownership rights to farmer tenants were not effective, because of a court ruling that after the initial process of land reforms, tenancy rights could not be conferred unless the tenants could produce a signed agreement with the owner as required by the 1964 Land Act. The number of landowner-cultivators thus arose in this period. Many cultivators had to surrender their rights because they could not present a written agreement. This happened to those who had made informal arrangements for sharecropping[32]. Over the decades, the number of recorded tenants declined[33] but informal tenancy (without documents to prove the rights of the tenants) in the form of sharecropping increased.

The farmers had developed a way of sharing the costs involved in irrigation between the land-owner and the cultivator when the area was irrigated by the kulo. Shallow groundwater irrigation and later deep tubewell irrigation (after the transfer of tubewells) presented a cash-intensive form of irrigation to the farmers over the labour intensive kulo irrigation to which they were used to. It is not known how cost sharing mechanisms between land-owner and cultivator have been affected by this change.

A study was conducted by Ravi Bhandari (1999) on land tenure and economic growth in the villages neighbouring the case study villages in this study. He showed that sharecropping was relatively efficient as compared to owner-operated farms. He attributed this to the new phenomenon of cost-sharing that had come up between tenants and landlords in all modern inputs. One of the case study villages in his study was rainfed while the other was irrigated by the kulo. In the villages in this study, farmers have moved from surface irrigation to groundwater and then further to conjunctive use. I therefore examine how cost sharing mechanisms have evolved as farmers switched over from labour intensive kulo to cash intensive groundwater. I do so in order to understand how could affect the way farmers made groundwater is managed.

The other issue that comes up prominently in the South Asian debate on agrarian structure is that of farm size and fragmentation, and its effect on productivity, especially its assumed negative impact on the emergence of collective action in the area of water control. Land fragmentation is ongoing in the Terai. It has been mentioned as one of the major constraints to the expansion of group shallow tubewells, despite a high level of subsidy that was offered for it (Gautam and Shrestha 1997; Koirala 1998). There is a subsequent subdivision of land into a number of smaller holdings. There are also a number of non-contiguous plots of land within a single holding. Out of the total 141367 hectares area of Rupandehi district, 85122 hectares have been demarcated as cultivable land. Of this, 27 percent of the landholders were small farmers who owned one or less than one hectare of land. Fifty-five percent owned land between one to four hectares and 18 percent owned more than four hectares. Land fragmentation is high with 18 percent of the farmers owning more than ten plots. Only 20 percent of total landholdings are held in one piece, and the majority of these are less than one hectare. Most of the farmers held plots in at least three and up to seven locations (DADO Rupandehi 2000). Land fragmentation has resulted both from the laws of inheritance as well as the continuous process of land transactions, as more and more settlers arrived and local Tharus started selling land. The value of land has increased remarkably after the initiation of the Bhairahawa Lumbini Groundwater Irrigation Project. Land size or fragmentation and land scatter are very important elements that shape the way choices can be made in irrigation, especially in groundwater.

Surface Irrigation in the Study Area

Numerous kulos exist in Rupandehi District[34]. The Sorha and Chattis Mauja irrigation systems combined form the largest farmer-constructed-and-managed irrigation system that is still functioning. Various studies[35] have dealt with organisation and management in Chattis Mauja by the farmers in detail, but I summarise some of the aspects of these kulo systems here. Two of the sites in this study are located in the tailend villages of these kulos, and therefore have a long historical and hydrological relationship with the larger kulo networks. Only one village in Madhaulia VDC was a member of the Chattis Mauja before it became part of the deep tubewell project. However, a substantial part of Tikuligarh VDC lies within the Sorha Mauja boundaries. The BLGWIP command area started from the tailend villages of these surface irrigation systems. Hence the project 'cut off' a portion of the established relationships between the villages, and introduced a completely new technological and institutional environment to these villages.

The Sorha Mauja and Chattis Mauja function as separate irrigation systems, but they also have a joint committee at the point where they share the water from the Tinau River. The area irrigated by the Sorha Mauja lies adjacent to the Tinau River, while Chattis Mauja irrigates an area that lies on the east side of the present day Butwal-Bhairahawa Highway. A government-constructed canal disrupted the flow into Sorha Mauja. The two systems have been sharing water from the Tinau River at a common intake since that incident[36]. After the kulos started sharing a common intake in the Tinau, they are collectively referred to as Sorha-Chattis Mauja. The farmers on the side of Chattis Mauja regard Sorha Mauja as its branch canal while the latter are adamant about their own 'individualism' and history.

Chattis Mauja means thirty-six revenue villages (birta) while Sorha Mauja means 16 revenue villages[37]. Even though the irrigation systems have been named as Chattis (36) and Sorha (16), the number of maujas irrigated by each is much higher than that suggested by their names. Chattis Mauja[38] and Sorha Mauja[39] each have a written constitution. The constitution of Chattis Mauja (1979 and amended version 1989) stated that the total number of villages irrigated by Chattis Mauja was fifty-four. The official records taken before 1994 indicate that Chattis Mauja irrigated 3500 hectares while Sorha Mauja irrigated 1500 hectares. The studies conducted Previous studies on Chattis Mauja did not include the area that was inside the command area of the BLGWIP. Eleven villages that were irrigating from the Chattis Mauja became part of the BLGWIP. The new constitution of 1995 denotes that there are sixty-two villages under the irrigation system. Likewise there has been an increase in irrigated area from 6750 to 10000 hectares. The 1991 constitution of Sorha Mauja states that it has 34 maujas under its jurisdiction. The official name of the irrigation system in the constitution is Sorha Mauja Sinchai Samiti (Presently 33 Mauja). There was no major revision in the constitution of this kulo in recent years. The Agricultural Development Office, Bhairahawa indicates that an area of 11110 hectares (DADO 2000) is irrigated by these surface sources.

History has it, that the actual work of constructing these kulos was initially started by the Tharus. Even though the lands in the area were given as grants to other people from outside the villages, it was the Tharus who managed the operation of the kulos. Yoder (1994) writes that the construction of the Chattis Mauja kulo was initiated by a local ruler but was managed by the Tharus until the early nineteen-fifties. The history of the canal can be traced back to the time of Prime Minister Janga Bahadur Rana (1846-1863). Similarly the history of the Sorha Mauja is also traced to a similar time frame. However, it is believed that these kulos existed before that time. The local people still debate the issue of which of the two kulos was constructed first.

History of these kulos has been that of struggles within individual systems as well as between the two kulo networks. These kulos were initially constructed to irrigate the areas inhabited by the Tharus. However, as migration increased into the districts, they expanded towards the north. The contest for water increased as more migrants came in to settle in the district after malaria eradication programme was implemented.

The area under both the kulos belonged under the jurisdiction of different jimidars (landlords). The local Tharu landlord, also called the Chaudhary acted as an intermediary tax collector and was also in charge of overall management of the kulos. The Tharu landlord therefore held absolute power over the farmers and all farmers had to participate in the construction of headworks and canal cleaning under compulsory labour. The system of jhara continued to exist as long as the older system of land tenure was in effect. Oral history states that there was compulsory labour for irrigation. When this was called, it was expected that all the households irrigating from the kulos had to contribute labour under any circumstance. The Chaudhary called the farmers for compulsory labour. Each farmer had to contribute labour irrespective of the size of land he cultivated. This labour was for making the headworks and for canal cleaning. The landlord and those loyal to him punished those people who disobeyed the call. It is said that the goats and cattle belonging to the farmers were either taken away or they were physically punished. The local landlords held absolute power over the tenants.

Studies in Chattis Mauja indicate that villages had already started opening up in the areas between Butwal and Bhairahawa as far back as 1945. A jimadar brought settlers from the hills and from India to clear land and open up cultivation. This led to the extension of the existing canal. An outlet was constructed in 1947 so as to irrigate the newly opened areas. The first major recorded changes to the irrigation system thus trace back to this period.

Similarly another incident in Sorha Mauja area took place around 1949. This was related to the conflict between the farmers who used the kulos and the farmers who had access to jharan water. The villages of Karaunjhia, Sakhuwani, Mangalapur, Tulsipur and Kotihawa and Kunjalapur got jharan water when the village upstream irrigated. The water that was available was sufficient for irrigation. The jharan arrived in these villages when the monsoon rains and field-to-field drainage from the villages upstream augmented the subsurface flow. These villages had to contribute labour for system maintenance in order to claim their rights to the jharan. The six maujas defied the call for labour contribution (*kulahi*) because they could still get this water, which arose naturally. The Chaudhary filed a case in the court against these villages. The people in these villages used their contacts at the higher level to put pressure on the Chaudhary. Parts of one village belonged to a Rana family who held very important posts in the government. The farmers in these villages made use of his power to disassociate them from the kulo system.

The overthrow of the Rana regime in 1951 led to the abolition of the birta and the jagir system of land tenure that had helped them to exercise their power in rural areas. The jagir was abolished in 1951 and the birta in 1959. The Ukhada Act and Rules were formulated in 1961. This was the first step towards land reform, which was implemented when power had been taken away from the Ranas and bequeathed to the Shah Kings. The comprehensive Land Reform Act was, however, put forward only in 1964, after a long power struggle between the king and the democratic forces[40]. The Land Reform Act of 1964, with some minor modifications, forms the basis of much of the current law regulating landlord-tenant relations. It imposed landholding ceilings on landowners and tenants, defined and stabilized tenure rights and set rent controls. The most important feature was that it secured tenancy rights to the people who tilled the land.

The high influx of migrants started in the 1960s after the malaria eradication programme. The physical boundaries of the irrigation systems started changing after more and more hill migrants started settling in these areas. There was shift of power from the older original inhabitants, as the hill migrant became the dominant group in the area. The migrants settled in the foothills and the northern parts of the district. Because of this, the irrigated area expanded in the northern parts of the originally irrigated areas. Slowly, the areas that the Tharus had been cultivating since a long time turned turned into tailend villages. The organisational structure for irrigation management transformed from landlord control into a very different structure as the new settlers started demanding rights to water. The evolution of the organisation and management of irrigation by the farmers in the Chattis Mauja Irrigation System is well known in areas of irrigation management in Nepal. It is one of the most frequently cited examples of local institution building in irrigation in the country[41]. Both the kulo networks have similar organisational structures but are unique in their own way.

In order to understand the way in which these kulos are managed, first of all it becomes important to understand the structure of these kulos systems as conceptualised by the farmers.

A village that receives a right to open a branch canal is considered to be a member of the irrigation system. So the main irrigation system consists of groups of villages that are members of it. At the branch canal level, the village organisations are in charge of how water is delivered. It is this level that interacts with the farmers on a day-to-day basis. Each village has its own village level committees (*kulo samiti*). This committee keeps a record of all the households that use water within its jurisdiction. The committee at the main irrigation system level is not involved with these details. There are other levels of organisation: the village level, the area level and the executive committee. In addition to this, the Sorha Mauja and Chattis Mauja have a joint committee at the point where they share water from a common point.

Water allocation used to take place in proportion to land ownership. But as the kulo network expanded considerably, the basis for water allocation changed from shares proportional to land to shares based on water demand and the ability to provide resources to the system. The system is now known as *kulara* and the call for resource mobilization is now termed as 'kulahi' rather than jhara. Kulara refers both to water entitlement from the main system as well as to the obligation to mobilise resources. A high labour requirement is a major feature of the kulo. Resource mobilisation takes place mainly in terms of labour needed to maintain the system and to keep it operational.

The Tinau River is characterised by frequent flash floods. This necessitates much labour in order to repair and maintain these kulos. This is the unique characteristic of these kulos. This is the reason why they have defined water rights in this way. Thus, a village that wishes to join the system has to specify the amount of kulara of water they want. They have to submit labour needed for the irrigation kulos in proportion to the kulara they demand. So a village has to ensure that it can provide a certain amount of labour before making its demand. Apart from being related to labour mobilization, the kulara is also related to the voting rights in the main committee of the irrigation system. One kulara entails four voting rights to the main committee.

Chattis Mauja requires a great amount of labour every year at the headworks. This includes the desilting of the main canal and the branch canals. Besides the regular cleaning operations, there is a call for double or triple kulara from every village level committee in periods of flooding.

Sorha Mauja and Chattis Mauja share labour for canal cleaning for a length of 3.4 km at the headend. The length of the main canal after the part to where the two divide the waters is 7.6 km. In a normal year, not much silt is deposited, around 800 persons work each day for main canal desilting alone. It takes an average of 10-12 days for the task to be completed. During an emergency the labour requirement is doubled or even tripled. Labour can be called any time of the monsoon. Desilting is done before the rains start. However, the floods often coincide with peak transplanting time or other times when the farmers are busy in their own fields. Therefore the farmers have to try hard to balance the labour between their fieldwork as well as for the irrigation system. Yoder (1994) observed that main canal desilting for 7.6 km of canal below the Sorha-Chattis Division required 7600 person-days and another 7,300 person-days in cleaning the branch canals and field canals. This was in the period when the system boundaries did not include the present groundwater irrigated area.

In order to meet the labour requirements, each landowner has to either go for canal cleaning operations by himself or pay someone to do it for him. Both Sorha and Chattis Maujas have made rules to facilitate participation in kulahi. Each women headed household or households with only elderly people are allowed to pay a flat rate of fee for a year. In 2004, the rate was NRs. 10000[42] per 0.68 hectares (1bigha).

The kulo chairman of each village is responsible for maintaining the kulahi at the village level as well as for making arrangements to send people for kulahi to the main canal. A village that is not able to maintain the kulahi, automatically loses its right to water. A kulo that takes one share of kulara has to send four people for labour works in the joint headworks of the Sorha and Chattis and also in the main canal. Each village level chairman (Mukhtiyar) is responsible for the fines that have to be paid to the main irrigation system by individual village level kulos. He is also the main person responsible for managing the kulo inside the village. He is in constant communication with the main committee of the irrigation systems. A messenger from the main committee informs the village level chairman about the dates for canal cleaning and time of release of water from the main canal to the individual village level kulos. Each village is allowed to define its own way of water allocation and distribution. The main committee does not interfere in these activities. The village level chairman is responsible for sending the people for canal cleaning and maintenance works in the joint headworks of the Chattis and the Sorha and in the main canal. This is done in proportion to the kulara share of water. There is a fine if a village is unable to mobilize labour. This fine is paid to the main irrigation committee. by the village kulo chairman on behalf of his village.

Migration and Settlement

The population in Rupandehi is very diverse in terms of its ethnic[43] composition due to the continuous process of migration into the district since many years. Migration has therefore been one of the major forces that have restructured social life in Rupandehi District and in the study areas in particular. The socioeconomic backgrounds of the migrants and the location where they settle along the water course gives rise to social groupings which can have a bearing on how water is managed. The use of surface sources for irrigation had already expanded in a particular way before the groundwater tubewells were installed in the area in the study area. In this study, I try to examine how the deep tubewells with their own conveyance structures have interacted with existing relationships that were established by surface irrigation in Tikuligarh and Madhaulia. In Mahuwari, I examine the relations between migration, surface irrigation and the choice of shallow groundwater irrigation. Migration may generate a high amount of external remittance that increases the purchasing capacity of some farmers for agricultural technology like shallow tubewells and tractors. It can also lead to a decrease or increase in the availability of labour for agriculture. Such factors have a bearing on the way farmers make their choices of crops and sources of water for irrigation.

A major wave of migration took place in this district after malaria eradication programme in 1951. The Tharus[44] were living in large parts of the malaria-infested area of the district long before other migrants came in. There are also accounts that settlements existed in certain parts of Rupandehi district in the Lumbini area that date back to the time of Gautam Buddha and the reign of the Sakyas. Old religious documents mention that in the sixth century B.C. Lord Buddha resolved the Rohini river water dispute between Shakyas and Kollyas in Lumbini area (Poudel 2000)[45]. Regmi (1971) cites that Kapilvastu[46] and Rupandehi were nearly as well cultivated as Bihar or Banaras at the time of political unification with Kathmandu. Parts of Rupandehi and Kapilvastu were united with present day Nepal two decades after unification of the country in 1768. The parts that were referred to as being cultivated must have been the southern parts of the district that adjoin with India. The town of Butwal in the foothills in Rupandehi District had already been established as an important trading route in the 1700s.

Migration took place in different patterns in the Terai before and after 1951. Before 1951, the rulers tried to increase their revenue and control by encouraging individuals to reclaim wasteland or forest land. They put forth a provision that any person who could reclaim land situated at a distance of more than a day's walk from existing settlements could do so. If labour and resources of local peasants were not sufficient, the person taking over responsibility was permitted to bring people from India for that purpose. This person was granted tax-exemption for ten years and one-tenth of the total reclaimed area as his birta. Having initially arrived as labourers, Indian migrants settled down in the area. The Rana government (1846-1951) initially encouraged immigration from India as pressure on the cultivated land had not yet become critical. The hillsmen could not be persuaded to live in the hot Terai. Concessions and facilities were provided to any Indian who wished to move into the Nepali territory along with his family. The migrant was given free allotment of agricultural land, in addition to a homestead and free supplies for constructing a hut. After having settled, a migrant could serve as a Jimidar, a tax collecting functionary for the ruling class. According to Regmi, purchase of land and jimmidari holdings by Indians were stopped in the Terai in 1920, but immigration was not banned.

Migration after 1951 has followed different patterns. Migration in the country is largely understood as taking place in two trends: permanent or lifetime migration and circular migration[47]. The first one refers to that situation, where migration is motivated by the search for more cultivable land and a better life as people move from highland to lowland. It also refers to migration that takes place within the national boundaries. Circular migration meanwhile refers to that which occurs both within the country and outside the national boundaries. This can be for a long duration or a shorter duration. Labour circulation is also known as seasonal migration and has been a major feature of livelihoods in rural Nepal (Rose and Scholz 1980) and it is also true for Rupandehi.

A large number of hill migrants came down to Rupandehi from the western mountains in search of better life and settled there permanently. Another wave of migration was of people who had gone out of the country in search of better livelihood opportunities but came back to settle down in the district. These were mostly families of ex-armymen who had joined the British and later Indian armies and those people who had migrated to Assam in India and Burma for better life. There had been a major out-migration from Nepal in 1916, in accordance with the agreement of the then ruler of Nepal with British India. Nepalese men from mountain and hill regions started emigrating for employment in British India. The men from these families serve or have served outside the country for a considerable number of years. The families of the army men have bought land in Rupandehi and have migrated from the hills. Some of them still retain their lands in the hills that they lease out to some kin or neighbours. In many cases, the women head the household while the men work abroad. Most of the people, who are either retired from the army or are still in the army, belong to the Gurung and Magar ethnic groups. The migrants to Burma (Myannmar) were also of two types. Some had gone there in the 1940s as British Gurkha troops to fight the Japanese in World War Two, while others had migrated in search of better life. In those days, passage to India was more accessible than to Kathmandu for the people living in the western mountains. Some families stayed back in Burma working as dairy farmers or gem miners, while many decided to come back to Nepal, after Ne Win came to power in Burma and their property was confiscated. Some escaped to Thailand[48]. The *Burmelis* as they are called locally are settled back in different parts of the Terai. However, the majority of those originally from Arghakhannchi, Gulmi, Baglung districts and other western mountains of Nepal, came back to settle in Rupandehi.

There is also a substantial movement of agricultural and other forms of labourers from one district inside the country to another[49]; from Nepal into India, as well as from India into Nepal. In his study on seasonal labour migration in Nepal, Gill (2003), states that the Punjab state in India is the major rural destination for Nepalese migrants, while Delhi is the most important urban one. There is also a flow of seasonal agricultural labour from Nepal into Haryana and Uttar Pradesh, in addition to Punjab where the Nepalese work force join with the other seasonal labour migrates from other parts of India. The Nepalese work force is also involved in other non-agricultural labour in India.

The flow of migration from India into Nepal mostly takes place from the eastern areas of Uttar Pradesh and Bihar. This labour force is employed in agriculture as well as in the non-agricultural sector. The Indian migrants who work in agricultural sector are mostly concentrated in the Terai of Nepal. However, those who work in other sectors are also concentrated in urban areas of the country. There is also a movement of labour from Nepal to Arabian countries and to Southeast Asia, especially Malaysia.

Concluding Remarks

In this chapter, I have laid down the history and context of intervention in the study area. The different technological, social, political, cultural and legal and hydrological contexts in which irrigation activities take place have been described. Technological intervention in groundwater irrigation entailed the introduction of a new technology to the farmers who were actively engaged in negotiating rights to surface water for irrigation. Through the discussion on the agrarian structure, migration and labour in the study area, this chapter provides a background to understand how these factors interplay with each other to shape choices between different sources of water for irrigation. In addition, the history of the evolution of surface irrigation and relations in production provides a basis to understand the social relations that emerge in the locality and how they influence water use practices.

Notes

1 For details on Tinau river system refer to Gyawali and Dixit (2000).

[2] Loans from the ADB and WB and Saudi Fund accounted for 59 percent of the total investments with 20 percent from government resources. The direct payments made to large projects like the Bhairahawa Lumbini Groundwater Irrigation Project (BLGWIP) and the Sunsari Morang, Narayani and Mahakali projects and different other studies were not reflected in the government budget. These are estimated to be around US$20 million. An amount of US$13 million was invested in the groundwater resource investigations in these three projects itself. For investment in irrigation see Shah and Singh (2001).

[3] Nepal became a multiparty parliamentary democracy and constitutional monarchy in 1990 after the overthrow of one party Panchayat regime that was started by King Mahendra in 1961. The Panchayat Constitution was based on a tiered system of village, city, district and zonal panchayats or councils.

[4] For discussions on these issues see http://www.cgisp.org.np/sub.html; Koirala (2001), Vokes (2001) and Siddiq (2001).

[5] The Hydrologic Investigations Projects under the USAID began the evaluation of the surface water potential in the country. This project helped to introduce a continuous data collection system for planning irrigation and hydropower projects. The set up of the Groundwater section of the Department of Hydrology and Meteorology was also assisted. Some 95 stream gauging and hydrologic stations were set up.

[6] This was carried out through a grant from the World Bank.

[7] Deep tubewells had been installed by the colonial government in the neighbouring Indian state of Uttar Pradesh by the 1930s. By 1960, half a million hectares of land in Uttar Pradesh were under public groundwater irrigation schemes. See Alberts (1997), Clay (1974), Dhawan (1982), Kahnert and Levine (1993), Pant and Rai (1985), Saxena and Singh (1988) on information on groundwater irrigation in the Indo-Gangetic Plains. In similar areas in West Bengal in India and Bangladesh see Boyce (1987).

[8] Other large irrigation projects: Sunsari-Morang Irrigation Project with developed area of 41,500 hectare cost around US$4000 per hectare; Narayani Zone Irrigation Project with developed area of 28300 hectare cost US$1777.

[9] The other groundwater irrigation projects have been Kailali Kanchanpur Groundwater Irrigation Project, Mahottari deep Tubewell, Sunsari-Morang deep Tubewell, Birgunj Groundwater, the groundwater irrigation programme of the Janakpur Agriculture Development Project, Sagarmatha Integrated Rural Development Project and the ILC/NISP.

[10] This is the area cited in Shah and Singh (2001). The BLGWIP Project Completion Report cites it as 20000 hectares.

[11] The details on the process of transfer (the project as a whole) have been drawn from Olin (1994), BLGWIP report (1999) and from field work through the farmers in the study area for individual experiences.

[12] See explanation on end note number three.

[13] The rate was fixed at NRs. 200 (US$ 9.50 at the official rate of exchange) per hectare of land in the command area. It was first levied in 1987/88. It was raised to NRs. 400 (US$ 11.50 at the official rate of exchange) per hectare in 1990/91. Olin (1994) states that 69% of the farmers paid that water tax.

[14] The fixed cost or demand charge was reduced from NRs. 60 per kw installed to NRs. 20 per kw installed and the cost per kwh consumed was raised from NRs. 0.95 to NRs. 1.40 (US$ 0.03).

[15] The farmers had demanded that the Nepal Electricity Authority (NEA) take care of the transformers so that they could just be clients of the NEA.

[16] From Tillson, 1985.

[17] This is based on the study, Hydrogeologic Technical Assistance to the Agricultural Development of Nepal. David Tillson (1985). Asian Development Bank/United Nations Development Programe.

[18] The five methods are the sludge boring method, bogi boring, manual rotary, thokuwa and dug wells. For details on different local drilling methods practiced in the Terai see Gautam and Shrestha (1997) or the shallow tubewell manual of the ADBN Nepali version.

[19] Flowing Artesian Well: When the top of a well in a confined aquifer is below the potentiometric surface, water will flow out of the well under pressure.

[20] These tubewells had started operation for a short time when the field work for this study was conducted in 2001/02. Information on use of two out of four of these tubewells was obtained for the short period that they were in operation. Durganagar and Tikuligarh (new well) were used for more than 300 hours in a year. The deep tubewell committees of the other two tubewells were not able to provide the log books. They also reported much lower use hours in comparison to Durganagar or Tikuligarh (new) well.

[21] The history of landownership in Nepal had been largely drawn from the works of Mahesh Chandra Regmi from his books: A Study in Nepali Economic History, 1768-1846 (1971); Landownership in Nepal (1976) and Thatched Huts and Stucco Palaces: Peasants and Landlords in 19th Century Nepal (1978). References were also drawn from Rajendra Pradhan (2000).

[22] The term is different from Zamindars of India, despite phonetic similarity. Refer to Regmi (1976: 106-107) for differences.

[23] The regulations of 1861 stipulated that cultivated lands, if fetching an income equal to 5 percent of the total tax assessment on the area under jurisdiction, were assigned as Jirayat for the Jimidar in East Terai, while it was 10 percent in the West Terai. But these were expanded by the Jimidars at the expense of the ordinary landholders.

[24] Regmi, 1976; www.msnepal.org, www.infoclub.com

[25] The Muluki Ain or the Law of the Land was based on Hindu religious texts and customary laws and was the main source of law between 1854 to 1951. For details refer to Regmi (1978) and Pradhan (2000: 39-70).

[26] Those people who had land grants under the birta tenure.

[27] According to Regmi, 1971, all castes and communities were expected to contribute jhara, however Brahmins were granted exemption in 1813.

[28] Raikar: land owned by state. Other categories of land tenure were derived from raikar tenure Regmi (1976).

[29] Different types of landless waged labourers exist in the hills and the Terai. In the haruwa system, the labourers do not take a loan as in some other systems. They are allocated a plot of land as part of the wage payment. The debt that the labourers incur within the contract period is done through the share of the harvest from this plot of land. There is however a condition attached to this system of labour. The family members, in particularly the wives of the labourers have to work for the same employer and are paid a fixed daily wage which is generally below the market wage rate (www.southasian.org, Shiva Sharma 1999 in www.antislavery.org)

[30] However, the practice of keeping the haruwa was quite prevalent in the 1970s and even till the early 1980s in Tikuligarh. Later on, the term was adopted by the migrants for paid daily wage labourers in agriculture. The haruwa were given around NRs. 5000 annually and around two quintals of rice. Field information, 2004.

[31] One of the most recent attempts towards land reforms was made through public announcement in the year 2001. The growing political unrest in the country led the government to announce these measures. The insurgency movement by the Communist Party of Nepal (Maoist) has raised the issue of land: land rights and land distribution as one of its major issues. The Deuba government announced measures to fix the ceiling on land in both the plains and the mountains. The land ceiling that had previously been fixed at 13.55 hectares for the Terai was reduced to 7.45 hectares. This set off different debates in political and policy making circles. A major concern among policy makers was how reduction of the ceiling in landownership would in the long run contribute to land fragmentation and be detrimental for agricultural productivity. According to the Land Tax Office of Rupandehi district, a total of 74134 hectares of land was registered in the name of 150860 people in 2001. There were 165 families who owned more than 8.2 hectares of land and held a total of 2150 hectares between them. There were 128 other families who possessed land between 6.8 and 8.2 hectares. If those families, who owned more than 8.2 hectares of land divided it within their family members (as per the laws of inheritance) there would be no more land for distribution.

[32] According to Yadav (1999), only about 300,000 tillers could be identified and granted tenancy certificates while the number of tenants had been estimated by the Agricultural Sample Census of 1961 to be more than 600000. The government acquired only 31841 hectares of land. Only 29123 hectares of land were redistributed to landless and small holders.

[33] Both the proportion of tenants and the area under tenancy declined from 40.4 percent to 9.5 percent and from 25.5 percent to 6.2 percent respectively over the period 1961 to 1981. The 1991 census however, showed an inconsistency.

[34] See Gyawali and Dixit (2000), for an overview of different water use

institutions in the Tinau River basin.

[35] Pradhan (1983), Stevens and Schiller (1993), and Yoder (1994), are the prominent studies. Others include those by Shrestha and Sharma (1987). Zwarteveen and Neupane (1996) examine women's participation in Chattis Mauja Irrigation System.

[36] Details on the merger of Chattis with Sorha see Yoder (1994).

[37] Historical accounts on Chattis Mauja have been drawn from Yoder (1994), Pradhan (1983), Shrestha and Sharma (1987), Oral Testimony of the ex-chairman of the CMIS, and field work. Historical accounts of the Sorha Mauja were derived through field work with key informants involved in the management of the irrigation system and the elderly.

[38] Chattis constitution, 1979, 1983, 1993, 1995 are the different written versions. The constitution was first written in the 1950s. It is also a member of the National Federation of Water User's Association Nepal (NFIWUAN). The NFIWUAN is a national level association which was established in 1999. It serves as a forum where all the registered Water User's Association that are scattered throughout the country voice their demands and involvement in policy making in irrigation.

[39] The Sorha Mauja irrigation system has a written constitution. The first constitution in written form was made in the year A.D. 1974 according to a chairman, who oversaw the management of the kulo during the early 1970s. The constitution was made after the general assembly meeting of 1991 and is still being used. It was revised in 1994. It is registered with the district administration as an irrigation system with an elected water users' association. It is also a member of the National Federation of Water User's Association Nepal (NFIWUAN).

[40] Political events at the end of the 1950s, were characterized by power struggles, between the democratic forces and the palace. In an attempt to quell the constant political turmoil, King Mahendra announced plans in 1958 to form the partyless government and to appoint a commission to draft a new constitution. The King banned the political parties and introduced the Panchayat Constitution from 1961.

[41] Uphoff (1986) has cited Chattis Mauja irrigation system as one of the four best models of irrigation system in the world.

[42] Exchange rate 1US$ was equivalent to NRs. 73.67 in 2004.

[43] There are more than sixty ethnic groups in Nepal, among which the Tibeto-Nepalese race and the Indo-Nepalese race are the major groups. A large number of these live in Rupandehi. Forty-three percent of the population in the district speak Bhojpuri. Bhojpuri is largely spoken by the Tharus as well as the Indian groups. Twenty-nine percent of the latter also speak Awadhi. Eighteen percent speak Nepali while one percent speaks Newari. The majority of the hill migrants speak Nepali but each ethnic group has a dialect of its own which is also in practice. There is a unique mix of people of all descents in Tikuligarh and Madhaulia VDC. For

better clarity and recent data on castes and ethnic groups in Nepal refer to Gurung (2003) on the social demography of Nepal.

[44] The Tharus were believed to the immune to malaria. According to Guneratne (1998), it is probable that they lived in these jungles because they could not establish themselves elsewhere. He cites colonial British accounts of Tharus that characterize them as timid and retiring in the face of more organized and aggressive people, abandoning their land and retiring deeper into the forest in face of encroachment. Bennett (1978), Cruickshank (1891), Stevenson-Moore (1900), Nevill (1905).

[45] Mention is also made of such an account in Benjamin (1994:27).

[46] Kapilvastu is the district adjoining Rupandehi to the west. Bihar is the north Indian state bordering the central and eastern Terai of Nepal and Banaras is a city in Uttar Pradesh state in India.

[47] Library of congress country studies: lcweb2.loc.gov/frd/cs/nptoc.html

[48] Nepali Times Issue #129. 24-30 January 2003.

[49] Labour for agriculture is met through by different means: daily wage labour, exchange labour (*perma*), contracting groups of seasonal labourers. Most of the smallholders in Tikuligarh and Madhaulia try to exchange labourers between families. This is mostly done for planting and harvesting operations. However, when this is not possible, they hire daily wage labourers. There is also a system in some villages of bringing in groups of labourers from the northern district of Dang. This is done through the contact amongst the Tharus in this district with the Tharu labourers from Dang through personal contact. This contract includes arranging transportation for the agricultural labourers and also arranging food and lodging for them. Seasonal variation in time of planting in different areas allows the labourers to arrange time for extra labour. There is also a flow of labourers from the eastern Terai district of Rautahat. Most of the people from Rautahat who come to find work in Rupandehi are from sugarcane growing areas of Rautahat and therefore find time to come for paddy cultivation in Rupandehi. Labour for surface irrigation is met through by the farmers themselves or sometimes by hiring daily wage labourers. Land preparation throughout the study sites is carried out by means of tractors.

3

Irrigation Water Use and Production Dynamics

"We do need groundwater also, but we find the soils have become harder since we started using the deep tubewells, we like to use the surface water as much as possible because it improves the quality of the soil"

-An old farmer in Madhaulia.

"We try to get the maximum out of these small plots of land, we have no other source of income and we find the pumpsets are very useful".

-A farmer in Mahuwari.

"We do not have a kulo anymore; I personally prefer the deep tubewell because it is cheaper than having a pumpset, which I cannot afford".

-A small landowner in Tikuligarh.

In this chapter, I discuss the different ways the farmers in the study area ensure water access and organise production around irrigation. In order to do so, I examine the historical changes in water use practices in the area, how farmers make use of different sources of water for irrigation, how they define property rights to them and the way they have developed and manage the different complexes of water. This study shows how tubewells structured the way farmers had to behave with respect to irrigation but also how farmers themselves worked out ways for irrigation. It shows how the design and the layout of the tubewells became enmeshed with the physical layout of the kulos and with the ecological and historical linkages that the different sources of water had with each other, to produce different institutions for managing water. The extent of use of different sources of water by selected farmers in the study area and the way farmers have been adjusting and making changes in the choice of crops and sources of water was also examined. These help in understanding of ensuing patterns of organization.

Irrigation in a Deep Tubewell Area

Tikuligarh

Tikuligarh VDC stretches over an area of 2176 hectares. The BLGWIP installed twelve deep tubewells in this VDC. The entire VDC was not part of the project at the same time. The northern and central parts of the VDC fell under the project area during the initial implementation phase of the project in the 1970s. The deep tubewells installed in these parts were functional from the early 1980s. The rest of the VDC obtained deep tubewells in the third phase of implementation of the project. The installation of these tubewells was completed in 1999. Officially, the whole VDC was thus part of the project. The total 'design command area' of all the deep tubewells inside this VDC was 1570 hectares. Thus 72 percent of the total area of Tikuligarh VDC is supposed to be under deep groundwater irrigation. The people in this VDC also use groundwater for drinking and household purposes. There were 1294 hand pumps of 1.5 inches diameter and nine open wells in the year 2002.

A first time visitor to Tikuligarh will get confused at the significant network of lined and unlined canals that criss-cross the area. The villagers point out that it is either the 'nahar' or the 'kulo'. They call the lined canals of the DTWs 'nahars' and the earthen canals that they have constructed 'kulos'. That means they associate the word 'nahar' (which in English means an irrigation canal) with an externally driven action.

Even though this VDC was provided with deep groundwater, the presence of the kulos and the use of jharan flow cannot be overlooked. The deep tubewell project boundaries overlapped with seven maujas of the Sorha Mauja network that lie in this VDC. Four maujas maintained their surface water rights throughout the project period. Besides this, a substantial area inside the VDC uses the jharans. Farmers have also installed shallow tubewells in certain parts of the VDC. These can be seen installed just next to the lined deep tubewell canals or jutting out from the middle of the fields.

The other thing that confuses the outsider is the boundaries of these maujas and the deep tubewells. At some points, the deep tubewell and its distribution system have cut off portions of a particular mauja. At some points, different parts of different maujas had to share the same deep tubewell.

Each VDC is divided into nine administrative wards. Each ward is made up of a village or a number of villages or settlements. In order to locate the different arrangements for irrigation in the VDC, I make use of the ward as a reference.

Water use in Tikuligarh is an intricate mix of villages connected to each other via the kulos and jharan, and connected or disconnected again by the deep tubewells. This interconnectedness is an important element that cannot be overlooked. The northernmost part of the VDC (Fig. 3.1), which is ward number five, used the jharan waters before the deep tubewell was installed. Even after the deep tubewell was installed they still used the jharan waters along with groundwater for irrigation. This village, along with the villages in ward number six, has developed its own method of managing the jharan waters.

Ward number six initially used the same deep tubewell as ward number five. However, this village has stopped using the deep tubewell four years ago. The deep tubewell is located in ward number five and the water has to travel a considerable distance before it reaches ward six. It was too expensive for the farmers in this village to take deep groundwater. There was also another reason for this village to leave the deep tubewell. The village is lowland and the water table is very high. Soil moisture content is thus very high throughout the year, except in the dry season. Some farmers have installed shallow tubewells in this village. This ward is again divided in terms of deep groundwater irrigation. Part of it uses the deep tubewell installed in ward number eight.

Ward number eight is a smaller ward than ward number seven or ward number four. However, even though it is smaller in size, it has two deep tubewells. Both deep tubewells were installed in the 'first phase' of the project. Both tubewells also fall in one village. This village is also a mauja of the Sorha Mauja.

In addition, ward number one, two and three and some parts of seven are also maujas of the Sorha. Wards number one and two belong to the 'first phase' area of the BLGWI project, while ward

FIGURE 3.1 Water use complexes in Tikuligarh VDC

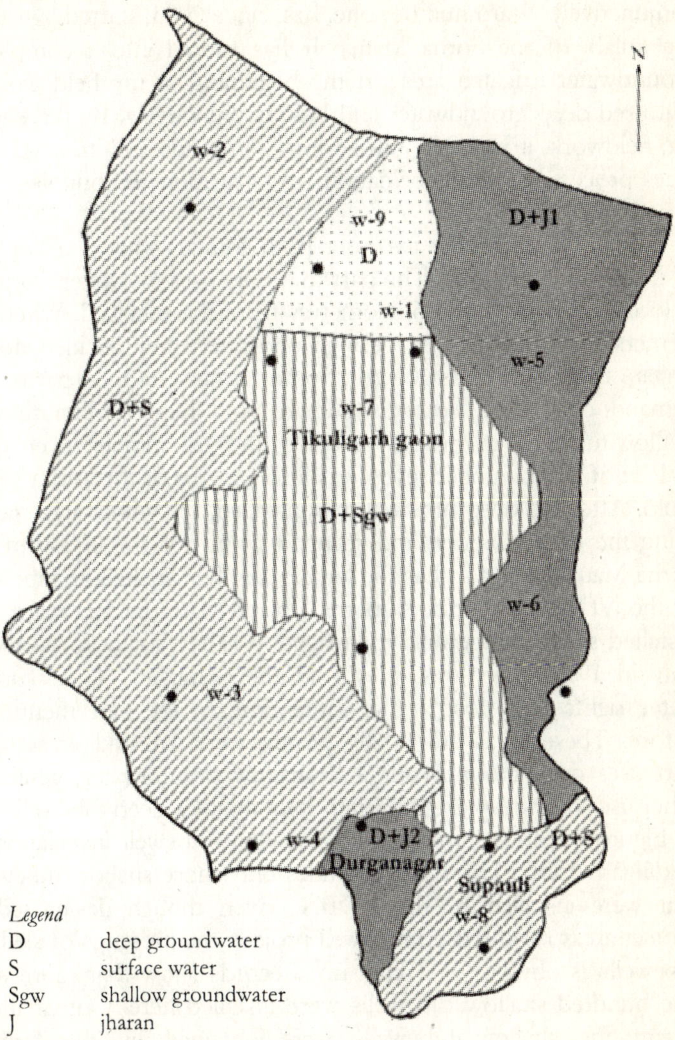

Legend

D	deep groundwater
S	surface water
Sgw	shallow groundwater
J	jharan
w	ward
.	location of deep tubewell
~	VDC boundary

Source: Field survey, 2001/04

three got its deep tubewell in the 'third phase'. Presently both ward number two and three use deep groundwater and kulo conjunctively. Ward number one, has, since 2003, started 'ignoring' the kulahi of the Sorha Mauja. It has turned into a completely groundwater irrigated area. At the beginning of my field work, it still used deep groundwater and kulo conjunctively. By the end of the fieldwork, it had fully turned into a groundwater-irrigated area. The process of transformation of irrigation is ongoing and dynamic.

Part of ward number two uses both sources while a major part of it uses the kulo only. The part that only uses the water from the kulo is located almost 1.5 kilometres from the tubewell. When the farmers were made to pay for groundwater, they decided not to irrigate from the deep tubewell. Some farmers stopped paying the demand charge of electricity. There is a high proliferation of shallow tubewells in this area as well as in ward number seven.

Part of ward number seven was also a mauja of the Sorha Mauja Kulo. After the deep tubewell was installed, the whole area started using the deep tubewell and stopped going for the kulahi of the Sorha Mauja. This part has the largest number of shallow tubewells in the VDC. The farmers in the upland areas in this ward had installed shallow tubewells for irrigation. They got deep tubewells only in 1999. After this, they sold the pumpsets. Their surface water rights were lost at the same time as the part mentioned before. These areas are totally groundwater-irrigated areas. One part uses deep groundwater and shallow groundwater, while the other uses only deep groundwater from the new deep tubewell.

Figure 3.2 shows the trend of shallow tubewell installation in Tikuligarh VDC. This figure includes only those shallow tubewells that were installed by the ADBN. Even though designated as project areas for the deep tubewell project, the presence of shallow tubewells is obvious in Tikuligarh. Records show that more than one hundred shallow tubewells were installed here. Out of these, twenty-one shallow tubewells were obtained by the farmers through subsidies offered by the Bank under the group shallow tubewell schemes.

A field count in the biggest ward in the VDC (ward number seven) shows that there are more than forty shallow tubewells installed. All these shallow tubewells are in use. There are even more shallow tubewells that the farmers have installed in their fields but are not being used. There are three deep tubewells in this ward. Two of the deep tubewells had been installed in the seventies and eighties.

FIGURE 3.2: Shallow tubewells installed in Tikuligarh VDC

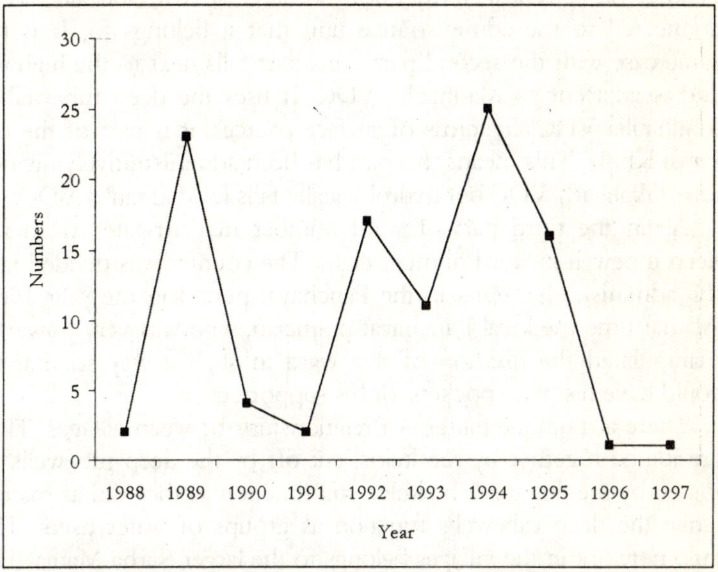

a) Source: Well Certification Records of ADBN, 1997
b) Source: Field Survey, 2002

Part of ward number four has its own subsurface spring source (jharan). In addition, the farmers in this ward used to augment this source of water with the field-to field drainage from ward number seven. When ward number seven stopped irrigation from the surface irrigation system, this source of water was no longer available for ward four. The farmers in seven used groundwater, which they started using more judiciously after the management transfer of the deep tubewells. The people in these villages had installed shallow tubewells in order to irrigate in the dry season. Part of this ward still irrigates from the surface irrigation system and drains. After the farmers had obtained new deep tubewells in 1999, many of them sold their pumpsets and are now using deep groundwater and jharan conjunctively.

It was very difficult to establish which area in Tikuligarh VDC is ward number nine. Ward number nine is made up of three segments; none of the three settlements are physically connected.

This is very rare because wards are not supposed to be physically disconnected. A part of ward number nine falls in the same location as the villages in ward number seven. Since it is located next to ward number seven, it is totally dependent on groundwater. Neither in physical terms nor socially or hydraulically, is it connected to the administrative unit that it belongs to. It is the same case with the second part. This part falls next to the highway and is adjacent to Madhaulia VDC. It uses the deep tubewell in Madhaulia VDC. In terms of surface sources, it is part of the old Lausi Khola. This means this part has been administratively divided into Tikuligarh VDC but hydrologically falls in Madhaulia VDC.

Again the third part of ward number nine irrigates from the deep tubewell in ward number eight. The country was divided into the administrative units in the Panchayat period in the year 1964. At that time the local Panchayat politician, who was very powerful manipulated the division of this ward in such a way so that he could have his 'vote pockets' or his supporters.

There is a unique pattern of relationships between villages. They are joined together by the kulos, cut off by the deep tubewells or joined by the jharans. The kulos still function as they did as maujas while the deep tubewells function as groups of water users. The kulo network in the villages belongs to the larger Sorha Mauja. The larger jharan network also stem from the Sorha Mauja. But when it comes to laying claims, none of the villages that use the jharan want to formalise their relationship with the kulo system.

The above description of the situation in Tikuligarh VDC shows that technologies, institutions and practices have changed as farmers have been making their choice of different sources of water for irrigation. Unique patterns of water use regimes emerged. These patterns or water use complexes are unique combinations of different sources of water. The different water use complexes functioning in this VDC are shown in the Table 3.1. The last column in this table shows the different entities that are present to manage the different resources in a particular complex. Each mauja and jharan is managed independently by a committee of water users'. The deep tubewells were designed to be managed by the water users group. Table 3.2 summarises the process of transformation in each ward as has been described in the above paragraphs. It shows the different stages of transformation of irrigation, as farmers made their choices between different sources of water.

TABLE 3.1: Water use complexes in Tikuligarh VDC

Ward No.	DTW Location	Surface Source	Complex	Organisation for Management		
				Dtw	Kulo	Jharan
1	Ward 1	SMIS	Deep groundwater	*		
2	Ward 2	SMIS	Deep groundwater	*	*	
3	Ward 3	SMIS	Deep groundwater & surface	*	*	
4	Ward 4	SMIS	Deep groundwater & surface	*	*	
5	Ward 5	Drains	Deep groundwater & jharan	*		*
6	Ward 5	Drains	Shallow groundwater & jharan			*
7	Ward 7 (1)*	SMIS	Deep groundwater and surface & surface	*	*	
	Ward 7 (2)	SMIS	Deep groundwater & surface	*	*	
	Ward 7 (3)	SMIS	Deep groundwater & shallow groundwater	*		
8	Ward 8 (2)	SMIS	Deep groundwater & surface	*	*	
9**	Ward 1	SMIS	Deep groundwater	*		
	Madhaulia	Lausi	Deep groundwater and surface	*	*	
	Ward 8	SMIS	Deep groundwater and surface	*	*	

**(1) (2) (3) indicate different tubewells in a ward*
*** Ward 9 is made up of three non-contiguous plots*

Source: Field Surveys 2001/2004

TABLE 3.2: The process of transformation of irrigation in Tikuligarh VDC

Ward No.	*First*	*Second*		*Third*	
1	Surface	Deep groundwater & surface	(1982)	Deep groundwater	(2003)
2	Surface	Deep groundwater & surface	(1982)		
3	Surface	Deep groundwater & surface	(1982)		
4	Surface	Deep groundwater & surface	(1999)		
	Jharan	Deep groundwater & Jharan	(1999)		
5	Member of Surface using Jharan	Deep groundwater & jharan	(1982)		
6	Jharan	Deep groundwater & jharan	(1982)	Shallow groundwater & jharan	(1992)
7	Surface	Deep groundwater & surface	(1982)	Surface (largely)	
				Deep groundwater & surface	(1992)
	Surface	Deep groundwater & surface	(1982)	Deep & shallow groundwater	(1992)
	Surface	Shallow groundwater		Deep groundwater	(1999)
8	Surface	Deep groundwater & surface	(1982)		
9	Surface	Deep groundwater & surface	(1982)	Deep groundwater	(1992)
	Jharan	Deep groundwater & jharan	(1982)		
	Surface	Deep groundwater & surface	(1982)		

Source: Field Survey 2001/2004

Madhaulia

Madhaulia VDC should literally be drowning in irrigation water. It spans a total area of only 1102 ha and is thus half the size of Tikuligarh VDC. Five deep tubewells were installed in this VDC. They were designed to cover an area of 596 hectares. A part of this VDC is also irrigated by a kulo from the Chattis Mauja. This kulo irrigates 200 hectares. Two different jharan systems also exist here.

One jharan system comes from Tikuligarh VDC and irrigates the northern part of the VDC. It irrigates an area of 66 hectares. Besides these two systems, there is a local perennial stream here. This stream called the Lausi, which irrigated a total of 456 hectares. Two kulos systems take off from this stream. Both the systems are independent of each other in term of management. A brushwood dam has been constructed at one point of the stream, this irrigates 136 hectares. A modern concrete dam has been built to irrigate 320 hectares.

There are also twenty shallow tubewells scattered across the VDC. A treadle pump was installed in 2003 for demonstration purposes. Altogether the potential to irrigate exceeds the total cultivable area of the VDC.

Groundwater is also used for drinking and household purposes in Madhaulia. These are mostly drawn from hand pumps. There are also free flowing artesian wells. Water from these wells are also used for domestic purposes. Altogether there are 101 free flowing artesian wells and 892 hand pumps in this VDC. The artesian wells flow continuously throughout the year with drop in discharge in the drier months.

Madhaulia, like Tikuligarh or any other VDC in the country, is divided into nine wards. A deep tubewell was installed in ward number six to irrigate ward numbers one, two, three and six. However ward numbers one, two and three no longer irrigate from this deep tubewell. Only the farmers of ward six use it. Initially, this deep tubewell covered the irrigation for 106 hectares. In 2001/2002, only the farmers living in the upland areas of this ward 6 used the deep tubewell, and the area irrigated by the deep tubewell thus reduced to a mere 6.8 hectares of land only. The people in ward number one always had access to the jharan called Sakhuwani, from Tikuligarh VDC. They reverted back to using the jharan instead of using the deep tubewell. At the same time,

FIGURE 3.3 Water use complexes in Madhaulia VDC

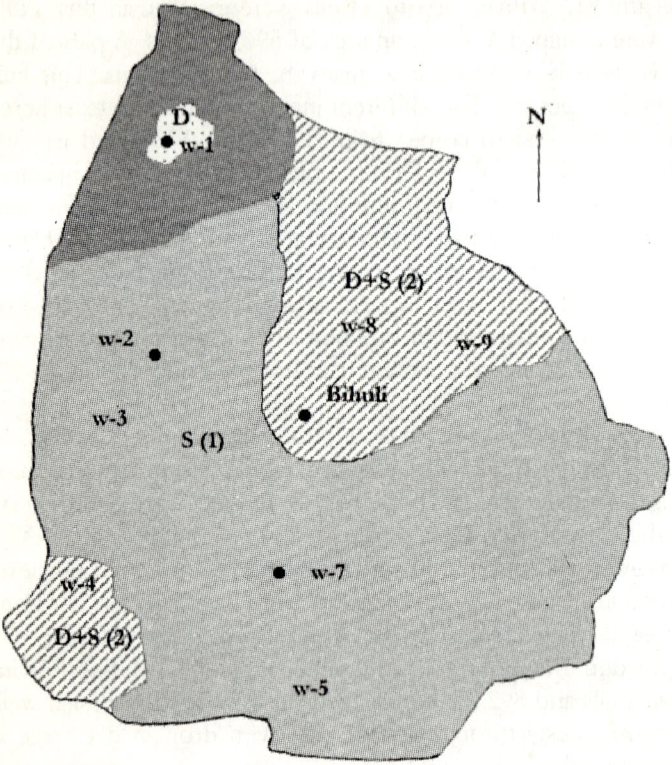

Legend

D	deep groundwater
S	surface water
J	jharan
w	ward
.	location of deep tubewell
	VDC boundary

Source: Field survey, 2001/04

the farmers in ward number two and three have also reverted back totally to their old kulo from the Lausi. They take water by means of a temporary dam in the Lausi Khola. One settlement is divided into two wards: six and one. Therefore, the area regarded as one settlement converted into two different complexes: a part that uses only deep groundwater while the other that uses jharan (Box 3.1).

The tubewell installed in ward number eight was designed to irrigate ward numbers seven and nine. The area has since decreased from 121 hectares to 41 hectares. This tubewell was installed just next to the river and was thus located in a lower level, while the area to be irrigated was upland. The people of ward number seven and nine have a new permanent dam across the Lausi further down from the temporary intake of ward numbers two and three. The tubewell in ward number seven was designed to irrigate ward numbers five and seven, but the farmers no longer use it. This area started using the water from the Lausi after the permanent dam was built. A part ward seven started using groundwater and kulo water conjunctively, while a part used only kulo water.

A deep tubewell installed in ward number nine was designed to irrigate ward number eight, which is the village of Bihuli. This is the focus village that is dealt with in the following sections. This deep tubewell now irrigates only 50% of the total designed command area. In this village the farmers have access to an old kulo from the Tinau. This village uses groundwater and surface water conjunctively. A part of ward number eight and nine also uses the deep tubewell in Gangolia VDC. Around twenty-seven persons pay money to this deep tubewell.

One tubewell was not taken over by the farmers. This area had turned into a residential area and the deep tubewell was abandoned. A small part of Madhaulia is also irrigated by a tubewell in Tikuligarh ward number nine. This is the part of the VDC that lies across the highway and merges into Tikuligarh VDC. This area is also irrigated by the same kulo system as the one that ward number two and three irrigate from. Figure 3.3 shows the location of the deep tubewells and the different complexes that are in use in the different wards in this VDC. Scattered around the VDC are shallow tubewells. These make up a more complex combination of water use from various sources.

There is a relatively low number of shallow tubewells in Madhaulia. Fig 3.4 shows the number of shallow tubewells that were installed by ADBN. One more shallow tubewell was installed

BOX 3.1 From conjunctive use to groundwater

The deep tubewell installed in ward number six was on the verge of closing down by the end of 2002. In January 2004, the Nepal Electricity Authority disconnected the power supply to the deep tubewell because the water users had not paid the electricity tariff for several months. This tubewell was designed to irrigate 108 hectares. In 2002, it was irrigating only eight hectares. By the end of 2003, the area further decreased to 6.8 hectares. Most of the farmers who irrigated from this tubewell in ward numbers one, two and three stopped using it after the transfer process. A portion inside the 'command area' of this deep tubewell had access to a jharan source that comes from Tikuligarh VDC. The other areas were irrigated by the local stream inside the VDC. Only the upland areas were still irrigating from the tubewell and paid the demand charge for it. Twenty households living in this area still need groundwater. They increased their share of payment of the demand charge in order to cope up with the electricity charges till 2003. In 2004, four out of twenty households stopped paying the demand charge. They formed a group and installed a shallow tubewell. The water users group increased the demand charge to NRs. 2206 per at the end of 2003. For several months they paid the electricity authority in instalment. In 2003, they had paid NRs. 7000 and an additional cost of NRs. 1,800. In the first quarter of 2004, they paid NRs. 16000 as electricity costs and additional cost of NRs. 4000. The farmers have to pay an extra charge of NRs. 300 as reconnection charge each time the power supply is cut off. There is also an additional charge of NRs. 5 for the application forms.

The farmers requested the ward leaders and the VDC to help them to clear the bills from the NEA. There were discussions as regards to whether it was proper to spend the budget allotted by the government to pay for the cost of electricity for the deep tubewells. The deep tubewells had been handed over to the water users group. The officials decided that this was a crucial part of village needs and allotted a sum of NRs. 10000 to this ward. It was the first time that the development budget was spent in this manner. However, this amount was still not enough to cover the total costs. This complex has moved from conjunctive use to isolation and use of individual shallow tubewells.

Source: Field Survey, 2001/2004

by the Groundwater Office under the DOI in 2001. The different water use complexes functioning in this VDC are shown in Table 3.3. The last column of this table shows the different entities that are present to manage the different resources in a particular complex. Each mauja, jharan and deep tubewell is managed independently by a committee of water users'. In addition the farmers in ward number six have a committee for the group shallow tubewell that they have installed. There are several dams on the Lausi and have thus been numbered accordingly. The first one is not yet in use and therefore, the others in use have been indicated as numbers two and three. Table 3.4 summarises the process of transformation in each ward as has been described in the above paragraphs. It shows the different stages of transformation of irrigation, as farmers made their choices between different sources of water at different points in time. The numbers in the brackets indicate the approximate year that the changes occurred.

FIGURE 3.4: Shallow tubewells in Madhaulia VDC

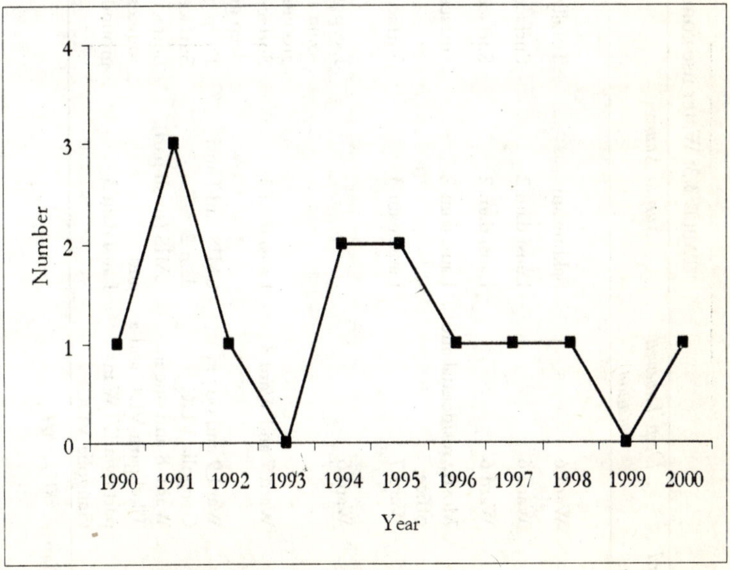

a) Source: Well Certification Records of ADBN, 1997
b) Source: Field survey 2002

TABLE 3.3: Water use complexes in Madhaulia VDC

Ward No.	*Deep Tubewell (location)*	*Surface Source*	*Complex*	*Organisation for Management*			
				DTW	Kulo	Jharan	STW
1	Ward 6	Sakhuwani	Totally jharan			*	
2	Ward 6	Lausi dam 2	Surface		*		
3	Ward 6	Lausi dam 2	Surface		*		
4	Mostly residential and office	Lausi dam 2	Surface		*		
5	Ward 7	Lausi dam 3	Surface		*		
6	Ward 6		Deep groundwater & some shallow groundwater	*			*
7	Ward 7 and Ward 8	Lausi dam 3	Surface with small part deep groundwater	*	*		
8	Ward 9 and one in Gangolia VDC	CMIS and Lausi dam 2	Deep groundwater & Surface	*	*		
9	Ward 8 and one in Tikuligarh VDC and a part by a DTW in Gangolia VDC	CMIS (small part) and Lausi dam 3	Surface from two sources & deep groundwater	* *	*		

Source: Field Survey 2001/2004

TABLE 3.4: The process of transformation of irrigation in Madhaulia VDC

Ward No.	*First*	*Second*	*Third*
1	Jharan	Deep groundwater & jharan (1982)	Jharan (1992)
2	Surface	Deep groundwater & surface (1982)	Surface (1992)
3	Surface	Deep groundwater & surface (1982)	Surface (1992)
4	Surface	Deep groundwater & surface (1982)	Surface (1992)
5	Surface	Deep groundwater & surface (1987)	Surface (2000)
6	Rainfed	Deep groundwater (1982)	Deep groundwater & some shallow groundwater (2003)
7	Surface	Deep groundwater & surface (1984)	Largely Surface with small part deep groundwater (2000)
8	Surface (two)	Deep groundwater (1985)	Deep groundwater & Surface (1993)
9	Surface (two)	Deep groundwater & surface (1985)	Surface from two sources & deep groundwater (1993)

Source: Field Survey 2001/2004

Irrigation in a Shallow Tubewell Area

Mahuwari

The village of Mahuwari lies near Tinau River and the Ghagra Nala. In the past, the source of surface water was the Mahuwari kulo. It originated in Chihiliya, in Chihiliya VDC. The jharans and the drains from the northern villages either drain into the Tinau or the waterlogged area in Chihiliya. This water had been harnessed by the farmers for irrigation. It irrigated the southern reaches of the Terai down to the border with India and beyond. Four maujas in the east of Mahuwari still irrigate from this kulo. They are Madhuwan, Sonaret, Chihiliya and Kouwa. However, Mahuwari, Bairia, Bairihawa, Bangain, Lahatorawa and Gargatti do not irrigate from this kulo system anymore.

All maujas were involved in labour contribution to the irrigation system when the landlords were in charge of managing the irrigation system. The majority of the villages on the tailend stopped contributing labour from 1981. A land registration programme was carried out this year. Even without contributing labour for system maintenance, drain water was still available in these villages. However, the flow of the drains has been disturbed due to an increase in settlement upstream. These villages have not received the drain water since the last six years. There had been an effort to rehabilitate the kulo system. This programme required the farmers to pay a certain amount of money for the process. These tailend villages did not pay the required money. Therefore, the farmers living in the headend did not include them in the irrigation system.

Mahuwari village in Hatti Bangain VDC, was also part of this kulo system. But the case of Mahuwari (ward number six) was different from other villages. Mahuwari as has been mentioned, is located next to the Tinau River and the Ghagra Nala. One of the major problems that it faces is that of river bank cutting. Heavy monsoon floods from the mountains cut off the river banks in the lower parts of the Terai every year. Around fourteen hectares of Mahuwari village was cut off by a flood. A portion of the *kulo* still exists in the northern part of the village. However, the *kulo* has now been converted into a temporary road. This portion still connects with the *kulo* in ward number nine. The inhabitants of the portion that was cut off moved to the other side of the river and established the present village of Mahuwari.

The ADBN records show that fifteen shallow tubewells were installed in Mahuwari (Figure 3.5). This includes both pumpsets as well as the tubewell. This village falls in ward number six of the Hatti Bangain VDC. A field count in the year 2002 showed that there were twenty-seven pumpsets in this village. The number of tubewells is almost double the number of pumpset. Three more flowing artesian wells and three shallow tubewells were drilled during the field research period. Mahuwari village has seven free flowing artesian wells. These are used both for irrigation and domestic use.

FIGURE: 3.5 Shallow tubewells with pumpsets installed by ADBN in Mahuwari village

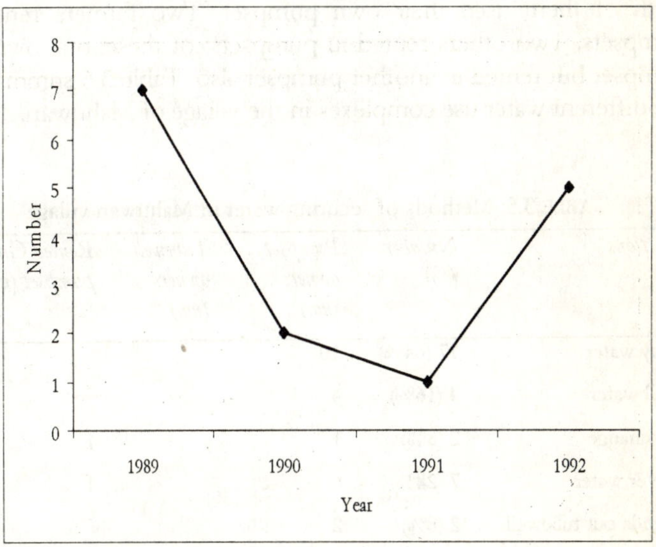

Source: Well Certification Records of ADBN, 1997

If the farmers in Mahuwari are to be classified in accordance to asset ownership in irrigation, they can be said to be of three types. There are those who own a complete shallow tubewell (with both pumpset and tubewell), those who own a tubewell only and those who own neither of the two. There are eighty-two households in the village. Out of this, sixty-seven families own some land, while others are landless. There are altogether twenty-seven pumpsets in the village. Forty landowners do not own a pumpset; that means sixty percent of those who own some land do not own a pumpset.

A random survey of twenty-five farmers in Mahuwari showed that seventeen out of twenty-five farmers bought water (See Table 3.5). Ten out of seventeen bought water even though they owned a pumpset. Four farmers sold water, while two farmers exchanged water between them. One farmer out of these two had his own pumpset. The second farmer did not have a private pumpset. He rented in a pumpset through a group scheme.

Farmers in Mahuwari also pump water from the river. These are mostly those farmers who own plots near the river. Seven farmers pump up river water for irrigation. Six out of these seven used their own pumpsets. The seventh one used a rented pumpset. Two farmers used the tubewells installed in their neighbour's plot (lent in) for irrigation. Two let their friends use their tubewells for free. Both of them used their own pumpset. Two farmers rent out pumpsets. Two others rented in pumpsets: of these, one owned a pumpset but rented in another pumpset also. Table 3.6 summarises the different water use complexes in the village of Mahuwari.

TABLE 3.5: Methods of securing water in Mahuwari village

Actions	*Number (%)*	*Pumpset owners (no.)*	*Tubewell owners (no.)*	*Rented Group pumpset (no.)*
Buy water	17 (64%)	10		
Sell water	4 (16%)	4		
Exchange	2 (8%)	1		1
River water	7 (28%)	6		1
Lends out tubewell	2 (8%)	2		
Lends in tubewell	2 (8%)	2		
Rent out pumpset	2 (8%)	2		
Rent in pumpset	2 (8%)		1	

Source: Field Survey, 2001/02/04

TABLE 3.6: Water use complexes in Mahuwari village

1	Shallow groundwater (bought)+(tubewell rented in)+river water
2	Shallow groundwater (own)+shallow gw water (exchange) +group owned
3	Shallow groundwater (own)+shallow gw water (exchange)+(bought)
4	Shallow groundwater (own)+(tubewell rented in)
5	Artesian (own)
6	Shallow groundwater (own)+(bought)
7	Artesian (own) +shallow groundwater (bought)
8	Shallow groundwater (own)+ shallow groundwater (exchange)
9	Shallow groundwater (own) + shallow groundwater (exchange)+ Artesian (own)+ River water pump
10	Shallow groundwater (bought)
11	Shallow groundwater (bought) +artesian bore (lend in)
12	Shallow groundwater (rents in pumpset from group)

Source: Field Survey, 2001/02/04

Cropping Pattern Choices and Water Use

The provision of deep groundwater largely reduced the dependence of the farmers on surface irrigation systems or on the jharans. The surface irrigation systems had initially been built to support paddy cultivation. Different factors led to an increase in the pressure on the surface water sources. Increase in migration and opening of the new canals had already led to an expansion in the surface irrigation networks. High-yielding varieties of paddy and wheat were introduced in 1967/68. The first improved varieties of wheat, (Lerma Roh-64) and (S-332) were introduced in the Terai and the mid-hills in 1968. Traditional varieties of paddy have been completely replaced by the high- yielding new varieties. There have also been shifts in the date of plantation between the older local varieties of paddy and the new varieties.

Paddy seedbeds are prepared through middle to late May. This is a dry period. Paddy seedlings are then transplanted around the third week of June. Sometimes the pre-monsoon showers contribute a substantial amount of rainfall in this period. However, this is one of the most important irrigation periods, when the fields have to be flooded with water. The monsoon starts in July and continues till September. In a year of normal rainfall the farmers do not need to irrigate in this period. Paddy is then harvested from mid-September through October. Wheat needs a crucial irrigation in the third week after it is sown. In the upland areas, the fields have to be made wet before sowing the wheat. In some low-lying areas the fields are still relatively moist even after the harvest of paddy. There are patches of low lying areas where only paddy can be grown. Subsequent irrigation requirements are met by the winter showers that fall in January. Farmers irrigate again only when the winters are very dry. Most farmers irrigated the wheat crop only once in the winter of 2001/02. Some did not irrigate it at all. There are other winter and dry season crops like lentils, chickpea and linseed that can be grown under totally rainfed conditions. However, the crops that are grown through the dry season in March, April and May have to be irrigated.

When the deep tubewells were installed, it provided the farmers with the opportunity for scheduling their irrigation and in making choices between different crops. Deep tubewells also made it possible to irrigate in the dry season. The cropping pattern in Tikuligarh and Madhaulia is largely rice-based as it is all over the Terai. There was no major 'diversification of crops' in large parts of Madhaulia. There have however, been some changes in the area covered by two new deep tubewells in Tikuligarh. The agricultural history of Mahuwari is very different with respect to these two areas. The farmers in Mahuwari have been following a more diversified cropping pattern. They have also been incorporating different combinations of crops from time to time.

There have been two distinct changes in the cropping pattern in the deep tubewell areas. One is the changes in the winter cropping in old deep tubewell areas after the transfer of tubewells. The farmers in Madhaulia, who have access to only deep tubewell or deep tubewell and kulo from the Chattis Mauja have either reduced their area under wheat cultivation or completely stopped growing wheat. Such a change can also be found in Tikuligarh. In Tikuligarh it is more common amongst smallholders in villages that do not have surface irrigation. The farmers have opted for rainfed crops like red lentils (masooro) in these areas. However this crop is very sensitive to water. In the winter of 2002, frequent rainfall destroyed most of the crop in the farmers' fields.

A calculation of the costs of production of paddy and wheat in 2001 showed that the farmers did not benefit profitably from either paddy cultivation or wheat cultivation (see table 5.1 and 5.2 in chapter five). Paddy is first allotted for household consumption for a year and is marketed only if a surplus remains. Yoder, in his study on Chattis Mauja (which did not include the study villages) in the year 1988/89, noted that farmers were reluctant to increase the area for cultivating wheat. He writes that high production costs compared to market value were factors identified by the farmers as determinants of acceptable risk. Without addressing those, changes in the physical system to make irrigation delivery more efficient are not likely to bring large increases in the area growing wheat (Yoder, 1994).

The farmers in this area also grow mustard and maize. Maize is grown in the dry season. Maize is mostly grown by the hill migrants as it is more of a hill crop. Only the farmers who are ready to irrigate from groundwater grow maize.

The yield of all major crops in the study area is high compared to the national or regional average. The average yield of paddy was 3.7 tons/ha in Madhaulia in 2001. This has always been much higher than that in Tikuligarh. It was 3 tons/ha in Tikuligarh while in Mahuwari it was 2 tons/ha for the same year. The heavier soils in Madhaulia area yield better than Tikuligarh. The average yield for paddy as recorded by the District Agriculture Office for the year 2000 was 3.6 tons/ha for the variety Radha-4 and Sabitri. After harvest, the majority of the farmers store paddy for household consumption for the coming year and sell the surplus. The average yield of wheat was 1.5 tons/ha in both Tikuligarh and Madhaulia in 2001. This is on par with the average yield in the country recorded in 1999 as 1.6 tons/ha. It is however less than that recorded for the district in the year 2000 which was 3.1 tons/ha. Mustard is irrigated once. However, it can be also be grown without irrigation. The yield of mustard was around 0.8 tons/ha across the sites in 2001. In 2001, the yield of mustard in Mahuwari was 2.2 tons/ha for those farmers who owned a pumpset and 1.5 tons/ha for those who did not. The average recorded for the district by the agriculture office was 0.72 tons/ha in 2000.

Figure 3.6 shows the monthly operating hours of a deep tubewell in Tikuligarh VDC from the year 1990 to 2001. The farmers who use this particular tubewell also irrigate from the kulo.

It is the last kulo to get water from the Sorha Mauja network. The farmers in this village had always retained their surface water rights even when they were getting free groundwater.

FIGURE 3.6 Monthly deep groundwater use from a tubewell where farmers also use the kulo for irrigation

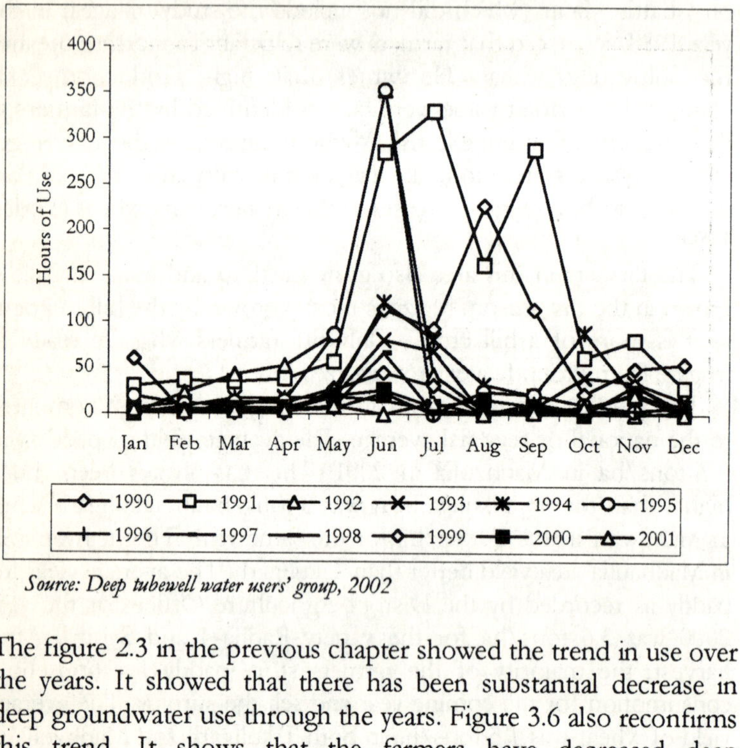

Source: Deep tubewell water users' group, 2002

The figure 2.3 in the previous chapter showed the trend in use over the years. It showed that there has been substantial decrease in deep groundwater use through the years. Figure 3.6 also reconfirms this trend. It shows that the farmers have decreased deep groundwater use throughout the year in all seasons. This tubewell area also gets surface water from the Sorha. The reduction in the use of groundwater for paddy shows that farmers have been using the surface sources more efficiently each year. There has not been a reduction in yield of either paddy or wheat according to the farmers. There has also been no change in major cropping pattern in this particular tubewell. The farmers in this tubewell area still plant paddy, wheat, mustard and maize and some vegetables for household consumption. They have a supply of surface water both for the paddy crop and the wheat crop. They are the last mauja of the Sorha and are the last to get its turn of surface water. It can be seen that there is now a much more efficient way of using both the sources of water. The prominent peak is in the pre-monsoon period. The sharp curves in 1991, 1995 and 1997 indicate the drought years. The increase in groundwater use is directly related to the low flow in the Tinau in these periods.

The choice of crops is also directly related to the design of the deep tubewells. The design of deep tubewells with an underground pipe flow has facilitated the farmers to follow a diversified cropping pattern especially for the cultivation of vegetables. However, not all farmers in all deep tubewells with this particular design are involved in vegetable cultivation. Only the farmers in two deep tubewells with this design have a more diversified cropping pattern. Box 3.2 gives the situation in one of these tubewell areas. In the previous paragraphs I had discussed how farmers have been reducing their area under wheat. The design of the tubewell did not facilitate the cultivation of other alternative crops like vegetables.

When a deep tubewell is used for one hour, the discharge is five times that obtained after operating the shallow tubewell. The operating cost for a deep tubewell is around NRs. 260 per hour. Operating a pumpset will cost the farmer NRs. 154 for five hours of operation (diesel price of 2002) if the engine consumes 1.25 litres per hour. If one was to make a comparison based only on the cost per hour, it is seen that operating the shallow tubewell is cheaper. However, the volume and therefore the depth of water obtained from deep tubewell is much higher. Farmers use the deep tubewell for irrigating wheat.

Wheat requires instant and fast watering and this is possible with the high discharge from the deep tubewell. The soil gets wet faster as the velocity of flow from the deep tubewell is also higher. They use the shallow tubewell to irrigate vegetables that are mostly used for personal consumption and sometimes to irrigate the paddy nursery.

Irrigation from Different Sources

Fourteen farmers were selected in the deep tubewell irrigated area, in order to understand how they irrigated for a complete cropping cycle. The twelve month period covered the winter crop of 2001 to the end of the monsoon season of 2002. The crop water requirement was calculated for these farmers for the specific cropping pattern that each farmer followed in that period. Only

BOX 3.2 Vegetable growers in deep tubewell areas

A majority of the farmers in Durganagar in Tikuligarh VDC grow vegetables. In the monsoon of 2001, vegetables were cultivated on 10 percent of the total area under the tubewell. This is a high percentage when compared to the older tubewells, where paddy occupied 99 percent of the area. In winter, wheat occupied 50 percent, mustard 45 and vegetables 0.03 percent. Farmers planted rainfed crops like red lentils in the rest of the area. In the new tubewell, vegetables occupied 13 percent of the area, mustard 28 and farmers grew wheat in 40 percent of the area in winter. The vegetables are grown for commercial purposes by the Lodhs and some Tharus in their own plot. The main crops grown here are paddy, wheat, potatoes, mustard and maize. Vegetables include beans, cabbage, cauliflower, tomato, radish, bottle gourd, bitter gourd, okra, and tomatoes (2001/2002 WUG, farmers). There are five families of Lodhs who live in this village. They belong to the Indian migrant group of the caste Lodh who are traditional cultivators. Most of the hill migrants are involved in other activities than only in agriculture. They give out a portion of their land for sharecropping to the Tharus. Thirty-five percent of hill migrants in this village give out land for sharecropping.

The sharecropper is responsible for marketing the produce. The landowner takes in the share of the produce. The produce is sold in the nearby towns of Butwal and Bhairahawa or in the local markets *(haats)*, set up in the town and inside the VDC. These markets take place a few days in a week at different locations. The farmers also sell their vegetables in the daily markets as well. The VDC also organizes a weekly market. The people in the VDC who do not grow vegetables come to purchase in these markets. Besides vegetables, poultry, goat meat, pork and fish are sold here. Sometimes wholesale vegetable sellers come to the village in their jeeps and collect the vegetables from the field itself. These vegetables are taken by these 'middle-men' to the western hills. Vegetable marketing has not yet developed into a cooperative like that in dairy farming. There is a chilling plant in Tikuligarh. Farmers deposit the fresh milk in the chilling plant in Tikuligarh, from where it is collected by the dairies and taken to Butwal. Vegetable farming is done only in larger scale on two new deep tubewell areas. Paddy is sold by the farmers to wholesale dealers who come into the village to weigh and buy the paddy and transport them.

Source: Field work, 2002

groundwater flow was measured. Surface water use was not measured. This is because these farmers are scattered across the study sites and they use water from jharans and kulos. The gap between the irrigation water requirement and the groundwater use is the amount that is estimated to be covered by the surface sources, the jharans, and the rainfall.

Figure 3.7 shows groundwater use and irrigation water requirements for thirteen farmers in the deep tubewell area. The fourteenth farmer in this area irrigates only from shallow tubewells. His water use is computed in the next figure (3.8) along with the farmers in the shallow tubewell area. The first six farmers irrigate from the new deep tubewells with pipe flow distribution system. The rest are located in the 'command area' of the old tubewells.

The first six farmers also have access to jharan sources. Farmers number seven to eleven do not have access to surface sources. The final two farmers are located in the kulo system. Eight out of these thirteen farmers in Figure 3.7 also have shallow tubewells.

Farmer number six sold the pumpset after the new deep tubewell had been installed. He irrigates totally from the deep tubewell and jharan. Three (farmer number one, two and five) of them still maintained their pumpsets so that they could use it in emergency. They use the deep tubewell and the jharan. Of these, farmer number five cultivates vegetables. She also sells the vegetables to make a living. She keeps the pumpset because she does not want to take the risk of not being able to irrigate when there is an electricity failure. All the first six farmers cultivate paddy, wheat, mustard, maize and vegetables.

Farmers number seven to ten are groundwater users. Two of them also own shallow tubewells. Farmer number seven grows paddy, wheat, mustard, maize and vegetables. He irrigates the vegetables using the shallow tubewell. The ninth farmer uses the deep tubewell and shallow tubewell for paddy.

Farmer number eleven uses both deep tubewell and shallow tubewell. She grows vegetables as a cash crop and irrigates them with the pumpset. Farmer number twelve has a plot in the upland and has not been able to irrigate from the deep tubewell. She depends only on the rainfall. She grows paddy and leaves the land fallow for the rest of the year.

Four farmers were selected from Mahuwari for the same purpose. The crop water requirements were calculated for a

FIGURE 3.7: Groundwater use and crop water requirements for selected farmers in Tikuligarh VDC

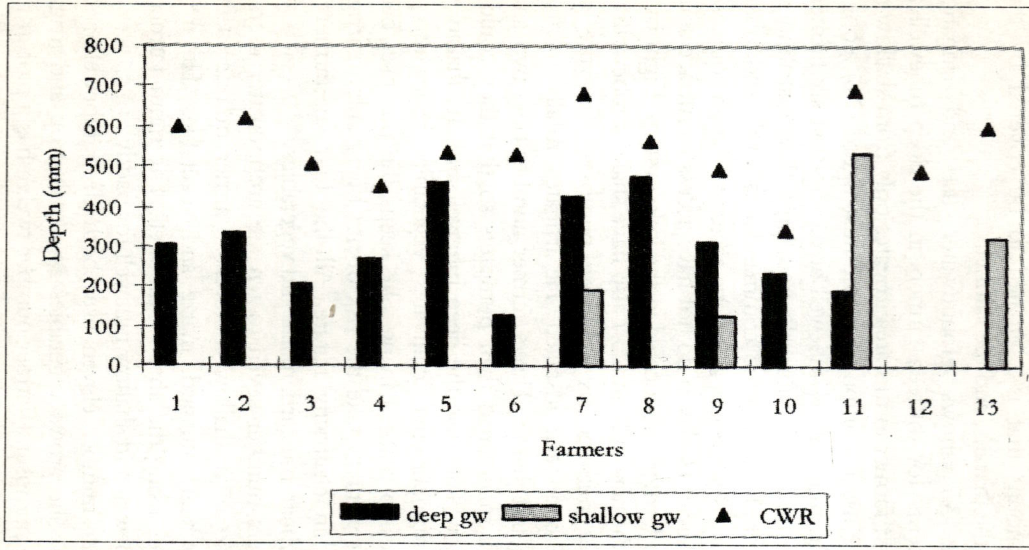

Source: Field measurements, 2001/02

FIGURE 3.8: Groundwater use and crop water requirements amongst shallow groundwater users in Tikuligarh VDC and Mahuwari village

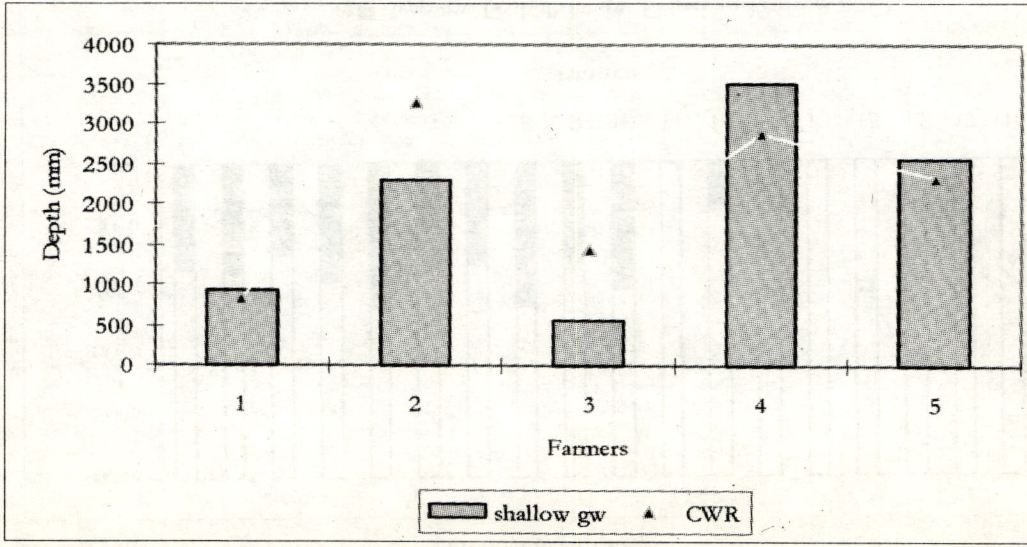

Source: Field measurements, 2001/02

FIGURE 3.9: Extent of use of each source of water by the farmers in Tikuligarh VDC and Mahuwari village

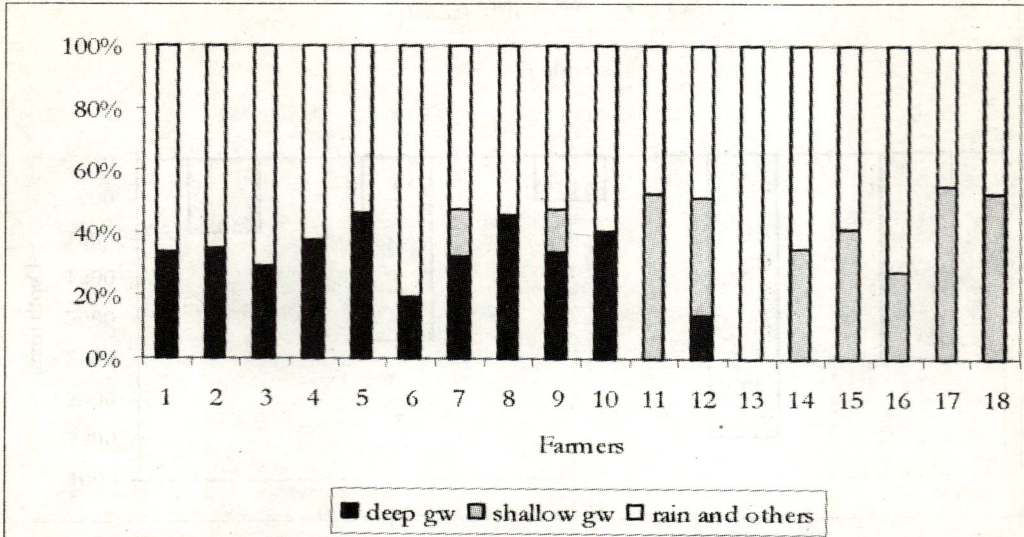

Source: Field measurements, 2001/02

BOX 3.3: Holding multiple sources of water in Tikuligarh VDC

Thake is represented as farmer number thirteen in Figure 3.7. He uses shallow groundwater and surface sources. He does not use DTW for irrigation but pays a certain amount of money for it. He reduced the area under wheat after he had to pay for deep groundwater. He irrigates dry season crops with shallow groundwater. He owns one hectare of land. In the monsoon he grows paddy and vegetables, while in the winter and summer he grows wheat, mustard, vegetables and potato. Thake paid only NRs. 150 (when the full rate was NRs. 442 per hectare per year) as demand charge in 2001/02. He spent NRs. 2000 worth of diesel for shallow tubewells and an equivalent of NRs. 2250, at the rate NRs. 100 per labour per day in one year for surface irrigation. The productivity of rice from his field was one of the highest in the whole area. He grew Sarju-52; the yield of which was 72 quintals per ha. Yield of wheat was 15 quintals per hectare. The total cost according to him was NRs. 3424 per year; which was the cost that he perceived. His total indirect costs were NRs 10640 that included depreciation cost, interest etc. His cost per cubic metre of water was NRs. 3.67. If the total amount that he spent in irrigation alone is calculated for a year (in terms of what he perceives his cost to be), he spent NRs. 150 on deep groundwater, NRs. 2250 for surface irrigation and NRs. 2000 for shallow groundwater.

Thake relies only on agriculture for his livelihood. He harvested 78 quintals of paddy. The market value for this was NRs. 46800 that year. However, it is much less than that because this does not account for the numerous harvest and postharvest losses. When his expenditure for irrigation is calculated in terms of paddy production, he spent: NRs. 150 equivalent to the production of his 0.0034 hectare of land for deep groundwater, NRs. 2250 equivalent to the production from 0.0476 hectare of land and NRs. 2250 equivalent to the production of 0.0476 hectare of land for surface irrigation. In one year he set aside around 0.092 hectare of his paddy field for paying for irrigation alone. That is equal to 6.5 quintals of paddy. This means he spent around 10 percent of his paddy production for paying the costs for irrigation alone. In other words, the production from 9.2 percent of his land is set aside for paying for irrigation. In fact it is more than this when the indirect costs are also included. Different inferences can be drawn from his actions: his partial involvement in the DTW affects the management of the DTW. The costs incurred in irrigation are high and the feeling of insecurity is also high even though he has access to three different sources of water.

Source: Field work, 2002

whole cropping pattern for each farmer. Figure 3.8 shows the shallow groundwater use of five farmers. The first farmer lives in the deep tubewell area but irrigates only from the shallow tubewell. He grows paddy, wheat, mustard and maize in this particular plot. The farmers in Mahuwari grow cash crops. The second, fourth and fifth farmers cultivate banana. The fifth farmer grows only cash crops: he grew banana, sugarcane and vegetable. He cultivates 0.34 hectare of land. Similarly, the second farmer cultivates 0.51 hectare of land. He grew paddy, wheat and mustard as well as had a small separate plot for sugarcane. The farmers in this village have did not cultivate sugarcane from the year 2004. The data shown in the graphs was taken for 2001/02. The banana plant takes more than fifteen months to mature and is harvested after that period. It is irrigated at an average of ten times in the whole cycle. It is irrigated in the dry season starting from April to the beginning of June, and then again from October to January if there is no rain. The third farmer follows the paddy-wheat cropping pattern with also potatoes and some vegetables. The fourth farmer cultivates paddy, wheat, mustard and banana in a plot of 1.52 hectares.

It can be seen that there is a tendency to over-irrigate amongst farmers who grow cash crop. The fourth and the fifth farmers have larger areas under cash crop. The gap between depth of use of groundwater and irrigation water requirements are fulfilled by the rainfall. Figure 3.9 sums up the extent of use of different sources of water by the same farmers in Figure 3.7 and 3.8 in Tikuligarh and Mahuwari. The total crop water requirement is taken as a whole and the percentage of use of each source compared. It can be seen from this graph that around forty percent of irrigation is done by using groundwater while the remaining sixty percent is fulfilled from surface and sub-surface sources, and rainfall. The total Potential Evapotranspiration (PET) for the entire period was 1394.7 mm and total rainfall was 1698.8 mm. This does not normally call for a large number of irrigations. However, the nature of rainfall is concentration in the monsoon season with sporadic rains spread across the rest of the months. The driest month was April, and there were seven rainy days even in the 'dry' month of May in this period.

Conclusions

An examination of the transformation of irrigation in the study area shows that there is a big difference between how deep groundwater irrigation was planned to be and the way it is. In terms of intervention and conceptualization by the agency, Tikuligarh and Madhaulia are deep tubewell irrigated areas. Field reality shows that they have transformed into areas with multiple complexes and thus diverse institutions for managing the different sources of water. Not all areas under the deep tubewells have fully converted to deep groundwater irrigation totally, even when they were supplied free water. The use of groundwater was high but the fact that some maujas still retained their water rights shows that the villages have been using groundwater and surface water conjunctively right from the start. The mauja-wise labour obligation rights were still maintained so that those maujas that did so, continued conjunctive use even after the transfer of wells. The maujas that were not able to do so, have converted into areas of groundwater use.

The farmers reduced their groundwater use after the transfer of tubewells. Field results show that the farmers fulfill 40 percent of their irrigation water requirements with groundwater in a year with normal rainfall.

An examination of the transformation of irrigation in the two VDCs shows that the ways in which farmcrs have been making their choice of different sources and institutions is a dynamic process. The complexes identified in the two VDCs are not yet stable complexes. The transitions outlined in Table 3.2 and 3.4, show that some villages incorporated deep groundwater and went in for conjunctive use of deep groundwater and surface water or jharan, some have either split off from surface sources or deep groundwater sources in another period of time. The farmers in the two VDCs had the options between three different sources of water: deep groundwater, surface water from kulos and jharan. Lack of legal restrictions on the use of shallow groundwater has made it easier for the farmers to develop it in the location they want to. There are, therefore, four options for the farmers in terms of sources of water in Tikuligarh and Madhaulia. In Mahuwari, shallow groundwater is the preferred choice amongst the farmers.

In order to understand the changes that irrigation systems undergo both in terms of boundaries and their performance, it is necessary to take into account these complexes of water use. For example if a whole mauja in a surface irrigation network leaves its water rights to surface irrigation, it affects the changes in the way labour obligations are defined by the larger irrigation system.

Similarly when farmers in a deep tubewell command area stop irrigating from the deep tubewells, this affects the way the committee in charge of managing the deep tubewells can manage it (examples cases cited in Box 3.1, Box 3.3).

The complexes in Tikuligarh and Madhaulia can be defined by the use of different sources. However, the complexes in Mahuwari have to be defined with respect to the ownership of the irrigation technology. This is because the definition of property rights to shallow tubewells is related to private ownership to technology. In Mahuwari most of the tubewells and pumpsets are individually owned.

In this chapter, I also examined how crop choice and choice of the source of water affects the type of social organization around water. The choices farmers make in terms of cropping pattern are influenced by ecology, characteristics of the irrigation technologies available, the market value of the crop, availability of labour and the extent of the farmers' dependence on agriculture. Each tubewell design helps to support particular cropping patterns. Related to the tubewell design is the economics of use of groundwater. The farmers who have access to both deep and shallow tubewells weigh the economics of using both tubewells. In addition, the shallow tubewells also work as an insurance against the deep tubewells. They are used when there is an electric power failure. Farmers cultivate cash crops only when they have assured labour supply. For example, a farmer needs year round labour if he or she ventures into vegetable cultivation. Those farmers, who have other sources of income, venture into vegetable farming through sharecropping. In old deep tubewell areas, the farmers install shallow tubewells for vegetables. Holding multiple sources of water also reflects the perception of the farmers of water insecurity.

Discussions on shallow groundwater complexes entail one more set of factors: land size and land fragmentation, and how these shape the social organization around water. The nature of shallow tubewell technology facilitates individual ownership. Therefore, the relationships that the farmers have to be involved in are more personal. The type of social organization that evolves comes through the interaction of all these factors. The complexes that come up are an outcome of the way farmers define their rights to the different sources of water and are embedded in all these factors. In the following chapters, I discuss how social relations shapes the way the different social arrangements coming into being around groundwater. In Chapter Four, I do this by examining the struggles of the farmers in the process of incorporating and adjusting to intervention in groundwater irrigation and their struggles over different sources of water.

4

Struggles in Conjunctive Use Complexes in Tikuligarh

"After twenty-eight years in the army, I retired as unit commander after the Kashmir war in 2001. I thought that taking the responsibility of managing the deep tubewell was nothing compared to the army. But now I feel that it will take me almost the same number of years to learn about the village and its politics, the government and the art of managing this tubewell"

-A chairman of a deep tubewell in Tikuligarh VDC

In this chapter I examine the struggles of the farmers in the process of adjusting to and incorporating intervention in groundwater irrigation in three different tubewell areas inside Tikuligarh VDC. In the first two cases, the deep tubewells are of similar design and were installed and managed by the project and then later handed over to the farmers. Both villages were also part of the same surface irrigation system, yet the two villages have come up with different ways of managing their water resources. Through the first two cases, I examine the reasons behind the different choices that the farmers make even when subjected to similar intervention processes. In the third case, the farmers obtained the deep tubewell through the 'demand based' approach. All three deep tubewells were ultimately managed by the farmers. In all cases, the role of the different actors and the strategies they employ, to gain control over groundwater and different sources of water is analysed. This chapter shows that the process of evolution of management around groundwater and its performance can only be understood by looking at how the deep tubewell technology has interacted with the history of relationships around the different sources of water in the area, the responses to ecological variability, the differences in power structure that existed in the villages before the intervention in groundwater irrigation, the agrarian structure and the shift in power from one group of farmers to another.

Deep Groundwater and Surface Water Use in Supauli

This case study is based in the village of Supauli, where the farmers use both deep groundwater and surface water for irrigation. This is the only village in the study area that made a joint rule regarding the use of both sources of water for irrigation. In order to understand this, the process by means of which the people in Supauli tried to incorporate intervention processes in groundwater irrigation and their struggle to manage both sources of water is examined. This is explained by examining the processes by which different actors in this village took the responsibility to formulating ways and means to manage the deep tubewells and the kulo. The role of key actors and their different strategies are analysed with respect to these processes.

The people of Supauli have always been very active in trying to increase the water delivery to their village. They maintained their surface water rights even when groundwater was free. The kulo is one of the oldest kulos in the Sorha Mauja network. Supauli is also the last mauja in the network and, therefore, the final village that gets it share of water from the irrigation system. The village used to take eight kulara from the Sorha Mauja. They later reduced it to two kulara and continued using the surface water even after the deep tubewells were installed.

This village is also unique in the sense that two deep tubewells were drilled here. One tubewell was installed in the north and the other in the southern part of the village. They were drilled in 1977 and 1979 and the installation was complete in 1982. The process of transfer of the tubewells started in 1992. There had been certain conditions set between the project and the water users' group during the process of transfer of the first generation tubewells. This entailed construction and rehabilitation of certain parts of the canals. This was performed in 1995/96. The water users' groups for both the tubewells were registered on 24 June 1997. The deep tubewells were then 'formally' handed over in July 1997 (BLGWIP 1999) after some rehabilitation works. These dates made the water

FIGURE 4.1 Deep tubewells and kulos in Supauli

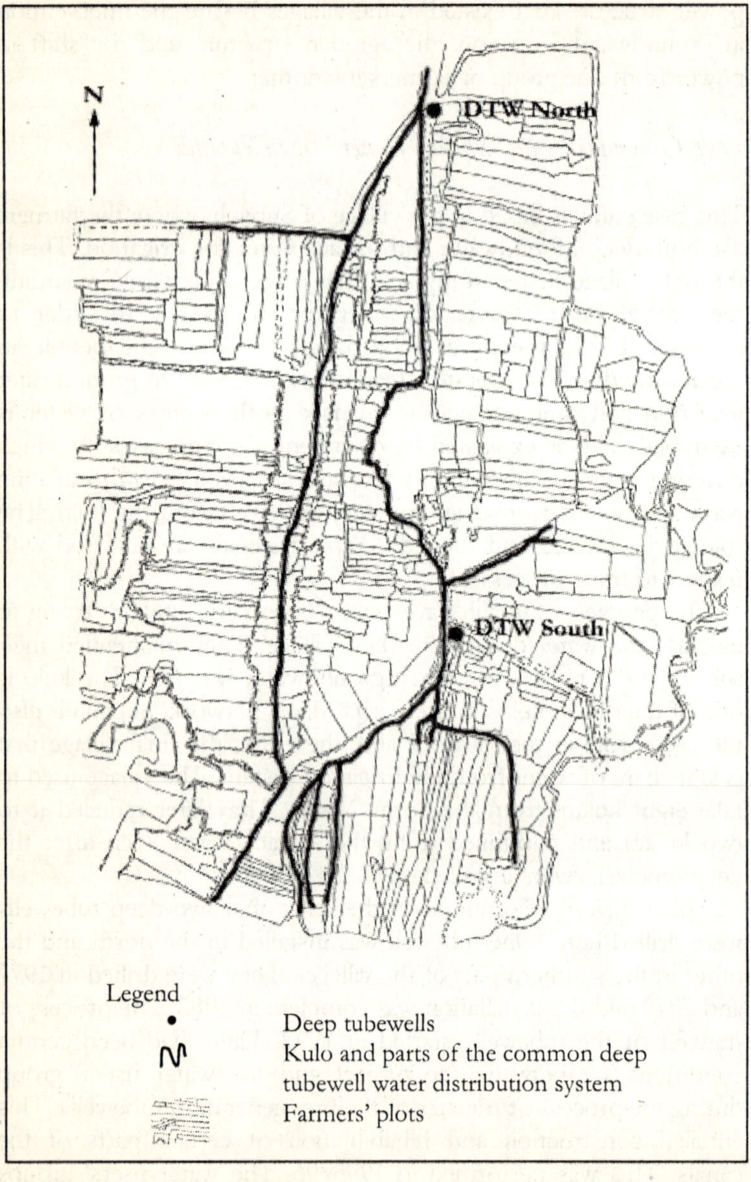

Source : Field survey, 2004

users' group a formal entity, but the farmers had been very active in making arrangements for irrigation long before this. The farmers of Supauli had been active in incorporating and coping with intervention in groundwater irrigation right from the beginning: before and after taking over the deep tubewells.

This village got two tubewells instead of one. This was because the landlord in the neighbouring village refused to allow the project to install a deep tubewell in his land. At that time, the project determined the location of the deep tubewells. The landlord was the Chaudhary who was also a local politician at that time. He was the Pradhan Pancha[1] in the Panchayat period. This landlord did not want his land to be subdivided by the permanent canals of the deep tubewells. The project had to find another location. Even if the installation had been driven purely by 'engineering standards', the project authorities still had to contact the farmers in order to obtain land for setting up the pump house. The Chaudhary's position as Pradhan Pancha, gave him the power to reject the tubewell.

The problem of the project was solved by the farmers in Supauli. The project was looking for a site and the farmers in Supauli were willing to contribute land for it. The deep tubewell pump house required an area of 0.15 hectares of land. The project compensated the farmers who contributed land for building the pump house. Two migrant families contributed land for the two deep tubewell pump houses. The villagers felt that they could do with two tubewells, especially since they were the last mauja to get surface water.

The role of Chaudhary has had a direct as well as indirect impact in bringing about transformation of irrigation in this VDC. First of all, his family was in charge of irrigation management activities in the area. They had been involved in construction as well as management of the irrigation system. They also held absolute power over it, before the migrants came into the area. The reason why he rejected the deep tubewell can also be understood in this light. Tharus have a deep historical relationship with kulos. The kulo was also part of Chaudhary's legacy. He had refused the tubewell even when it was known that the water was supplied free of cost. On the other hand, when he rejected the deep tubewell, farmers living in a part of the upland area adjoining his land did not get irrigation. Chaudhary could afford a shallow tubewell, while these people in the upland areas could not. His actions and decisions were not only linked to Supauli, but it also influenced the water use practices in two other cases study areas illustrated in this chapter: Tikuligarh gaon and Durganagar.

The migrants themselves were trying to establish their links with the higher political and administrative authorities. The people living Supauli were more active than the people who lived in the uplands. The latter were mostly very poor peasants, who had cleared the forest land and settled there. One of the oldest migrants still lives in the village and gave me an account of how he had 'brought in development' through his contacts with different projects and agencies in Bhairahawa. Developing contacts with projects and government offices was the way to build up a power basis. The migrants were more literate than the Tharus. They were more mobile because many of them had travelled long distance in-country or out of the country before they settled in here. They were also more vocal because most of them spoke Nepali, while the Tharus are more timid and spoke the local dialect. Some of the migrants used these opportunities and built up their political strength in the villages. They were what Guillet (1992) calls 'classic political entrepreneurs'. This sort of behaviour is common in all villages in this VDC, and not just specific to this particular village.

When the Panchayat regime was in absolute power from 1962 to 1990, most of the political parties were banned or they had gone 'underground'. After 1990, local-level politicians or actors who supported them held the strings of power. It is their 'connections' to the different political parties that gives them certain power positions inside the village. The local politicians in the VDC are made up of individuals from different sections of society. They are larger landowners with land holdings of more then three hectares, and also small landowners.

Search for alternative sources

After the transfer of the deep tubewells, the farmers of in Supauli requested an upstream village for a share of the jharan waters that were being used by that village. Supauli was not granted rights to this water. The jharan is part of the Sorha Mauja. The constitution of Sorha Mauja grants formal rights over the jharan waters to any mauja that contributes labour for the irrigation system. However, the upstream village has more power to make decisions over the jharan. This is because all villages that had access to the jharan waters had 'officially' cut off their ties with the Sorha Mauja in 1947 (see chapter two). These upstream villages were afraid that they would be argued back into the Sorha Mauja irrigation system if they gave rights to a formal member of the irrigation system like Supauli. The latter, is one of the oldest members of the irrigation system and had always maintained its water rights. Starting a new relationship with the old irrigation system was not favourable because these villages were already getting jharan water without having to make any contributions to the main irrigation system.

Struggles to retain the deep tubewells and the kulo

The tubewells came as a boon but turned into a burden for the farmers after handover. The leadership that emerged for managing the tubewells had to take care of all social complications that affect the different choices made by each and every farmer. Therefore, the major challenge for those who work in the executive of the deep tubewell water users' group has been the difficulty to convince people to pay for something that they themselves are not willing to pay for. The other difficulty for them was in finding out ways to enforce the rules and regulations for those who avoided payment. When the project was managing the deep tubewells, the arena was basically the project versus the farmers. Now it was the village and fellow villagers who took up the task of managing the deep tubewells. It was not possible to carry on with groundwater irrigation unless fellow villagers were made to pay. The leadership had to be capable of collecting the money for both the running cost as well as for flat rate of electricity, and for managing the deep tubewells.

The rate at which each farmer has to pay the demand charge is set by the executive committee of the water users' group of each tubewell. For example, both deep tubewells have a 75kW capacity. In 2004, each tubewell had to pay NRs. 2000 per month as demand charge to the NEA. The executive committee fixes the rate of demand charge, based on the payment it has to make to the NEA for a full year. This is then divided by the area that is being irrigated. Presently, only the owners of 116 hectares of land pay the demand charge for the two deep tubewells. This is only 50 percent of what should actually be collected in view of the design. The two deep tubewells in Supauli were designed to irrigate a total of 227 hectares of land. This included the whole of Supauli and parts of other adjoining villages. The area of Supauli is 169 hectares. Only two families from one of the villages outside Supauli still irrigate from the deep tubewell. The remaining command areas that lie in another VDC are not irrigated anymore. The canal that serves these villages is broken. Moreover, parts of the village lie in Madhaulia VDC. The farmers living in these areas are not included by the deep tubewell water users' groups. The farmers of Supauli claim that the farmers of these villages did not cooperate with them when the water users had been negotiating with the project regarding certain improvements to be made on the deep tubewell system. This had reduced the area under the tubewell. Presently, the tubewells and the kulo irrigate the area inside Supauli. The kulo irrigates 169 hectares, which is also the area of the village.

Several people figure prominently in the way they have taken over the charge of securing water for irrigation in the village after handover. Hari Bahadur took over the management of the north tubewell while Tek Bahadur took over the management of the south tubewell. Both of them, like the majority of the migrants in this village are from the western mountain districts of Parbat and Myagdi. They settled in this old Tharu village and presently, the migrant population and the Tharus are almost equal in numbers. Hari Bahadur is one of the early migrants to settle in this village.

Tek Bahadur came to the village in 1979, as a young army retiree, in his late twenties. Both Hari and Tek were involved in farming and also took part in various social activities in the village. They were both involved in the management of the kulo at different times. After the end of the Panchayat regime, they declared their allegiance to the United Marxist Lennist (UML) party. Hari was elected to the post of ward chairman in the first democratic elections in 1991. He also became the chairman of the north deep tubewell. There was another general election in 1995. Tek was elected to the post of ward member in this election. Another person from the same political party became the ward chairman and the chairman of the other deep tubewell. Tek carried on his job as chairman of the kulo and the deep tubewell.

According to Tek, there was increased pressure on the kulo, after handover. The farmers were also reluctant to pay the demand charge for groundwater. Moreover, many farmers had not been seriously taking part in the canal cleaning operations. A large part of the deep tubewell canal had been built over the existing earthern distribution system of the kulo network. The farmers blame the project for doing so[2]. However, some farmers had also actively taken part in this process. They were not willing to contribute extra land for two different distribution systems: one for deep groundwater and the other for the kulo. This was especially difficult for those farmers who owned smaller plots. The farmers who had lined canals in their plots refused to take part in canal cleaning operations which are part of the requirements of the kulo system. The managers faced challenges from both fronts.

The transfer of tubewells was made to the water users' groups for each well. However, directly or indirectly, the issue of provision of water, whether drinking water or irrigation water, is an issue for the authorities in charge of the village administration. A crisis in either one of these automatically becomes the responsibility of the elected ward members also. As an elected member of this ward, it was the duty of Hari Bahadur to oversee all developmental activities within the ward. In addition, he was also in charge of one of the deep tubewells. Tek Bahadur was in charge of the kulo and the other deep tubewell. They realised that it was necessary to put some form of control over both groundwater and surface water. A village meeting was called to discuss matters related to irrigation. They had already worked out their agenda. A condition was put forth in front of fellow villagers in the meeting held in 1993: 'any farmer, who wants to use the kulo, has also to pay the demand charge for the deep tubewell'.

This rule was put in effect, after it was endorsed by the majority of the villagers. This rule did not hamper the water distribution process because the farmers use groundwater in the months of May and June and also in the following months if the monsoon is late. Surface water is available in this village from July through September. The heavy monsoon rains in that period contribute a large part of the crop water requirements. Groundwater is used again in the winter months if the village does not get surface water in time or it is scarce. The two deep tubewells in this village, like all other deep tubewells installed by the project, were designed as separate units. However, in Supauli, the two deep tubewells are linked up by the old kulo network, so that, if there is a crisis in the south, the users are confident that they can distribute water from the north tubewell.

The leaders in this village were able to enforce the rule for joint use for several reasons. First and foremost, the boundaries of the deep tubewells, the kulo, the village and the ward coincide here. It is quite unique. This rule could not be made through the platform of the deep tubewell committee alone. A meeting of the deep tubewells alone could not be synonymous with the village. A meeting of the village and a meeting of the kulo is synonymous. Settlements initially developed around the surface sources. Historically the kulo has bound social life in both sides of the villages together. Besides its irrigation function, the kulo committee is also responsible for making arrangements for religious activity of the Siban puja[3], and for the social activity of road cleaning operations (sadkaai[4]). Both the northern and southern areas have been doing this together as a village. It was necessary for the whole village to be together to formalise the rules and regulations as regards water management.

It was easier for the leaders in this village to enforce the rule because of the more egalitarian landownership pattern in the village. The process of handover had been resisted by farmers who had large land holdings because they had to pay a large sum of money as demand charge. Seventy-four percent of the farmers own less than a hectare of land in this village. Nineteen percent of them own between one to 1.4[5] hectares, while five percent own 1.4 to 3.4 hectares. Only 1.7 percent own more than 3.4 hectares of land. This is basically the property of one family that lives in the village. The property is divided amongst the various family members. This family, according to the deep tubewell committee members, has been paying their dues and did not pose a problem. The majority of the family members live outside the country, while only one family member manages the farming. The leaders involved in decision making are also medium and small farmers. The majority of the farmers are owner cultivators.

Evolution of deep tubewell committees and their challenges

Enforcing the rule for joint use was part of the process of gaining control over both surface and ground water resources. The challenge of keeping the deep tubewells in running condition, and devising rules and regulations, and enforcing them is the biggest challenge for those who have taken up the responsibility to do so. The choice to use a common property like the deep tubewell or the kulo depends a lot on the people who decide to do so or are entrusted with the authority to do so. The process by means of which the committee for managing the deep tubewells has been evolving and the strategies that they have been working out for deep groundwater irrigation show how they have been striving to give continuity to deep groundwater irrigation, despite the odds that the technology imposes upon them. The transfer programme gave the water users' committee the legitimate authority to make and implement decisions and also to enforce the different rules and regulations with respect to deep tubewell management.

The formal set up of the committee for managing the deep tubewell is made up of seven members and a pump operator. The functions are basically carried out by the chairperson, secretary, treasurer and the pump operator. These are the people who are really involved in the day to day working of the tubewell. All members are not equally involved. There have never been elections for the committee. Taking up the responsibility of the deep tubewell, is, according to the present chairman of the north tubewell, like *'Kauso bhidauney'*. This means, forcefully putting or covering a person's body with thorns. (Kauso is a type of prickly plant). It is very difficult to find anyone who is willing to take over the charge of the deep tubewells. The meeting of the water users' groups is called in the village. People suggest the names of candidates they would like as chairman. It is endorsed after a majority of the farmers have indicated their agreement by clapping. The newly elected chairman also gets an opportunity to make suggestions for the people he would like to work with. The chairmen state that they are well aware of 'incorporating people from all ethnic groups (and gender) and different sections of society'[6].

The deep tubewell committee is one of the platforms amongst many, that aspiring local political leaders make use of. All development activities are carried out by creating certain 'groups'. Each farmer is a member of different groups at the same time[7]. Aspiring or elected politicians are very active members or leaders of such groups. The committee for managing the south tubewell changed only once after handover. A second committee was formed in 2003. A third committee was in effect in the north tubewell in 2004. The transition from one committee to the other in the south tubewell was smoother then in the north tubewell. In the south tubewell, the agreement to take up appointment came willingly from a young person: Thapa volunteered to take up the responsibility. He left his job as an agricultural technician in the government and came back to live in the village. He had come back to 'take part in the development of the village', and to build up a base for himself as a local politician. He also has an advantage to the rest of the people in his community. Most villagers have connections with the military. He has connections with the bureaucracy. This is an asset for the community. However, the new committee decided to keep the older chairman in a new capacity as treasurer of the committee. They also 'wanted to learn management strategies from him'.

The new chairman of the north tubewell came to live in the village after having spent twenty-eight years in the Indian army. He took retirement after the Kargil War in Kashmir, in 2001. Shyam Bahadur's family had been living in the village, but for him life in the village was a totally new experience. He had never lived in the village for a long time. Both new chairpersons support a newly formed political party, the Rashtriya Samata Party[8]. All leaders, old and new, share a strong kinship relation. However, the emergence of the new party on the local scene is something not to be taken amiss. Both chair persons had approached the leader of this party, who had been a minister at that time, to request for scrapping the demand charge.

Both new chairpersons have informal discussions with the older chairman and take advice from him on the management of the deep tubewells. I was also part of these discussions on several occasions. The three were well aware of the problems in each deep tubewells and also discussed ways of punishing those farmers who had not been paying their dues for the tubewells in time. Both of them were learning from Tek Bahadur. Thapa had woven himself right into taking part in the different activities in the village. He was made chairperson of the deep tubewell just before the BLGWIP had offered a budget for lining the deep tubewell canals that very year. With Tek's help he had negotiated with the project and had procured the fund for this lining. The project had set up a budget for renovation of ten deep tubewells to be spent that year, based on demand. Only those tubewells that were clever enough to contact the project on time were able to get the money. In addition, Thapa is also actively involved in a committee that oversees the rehabilitation of a temple in the area. The chairman of the kulo oversees the siban puja but Thapa organises the Durga Puja[9] at the temple[10].

There is a difference in the way responsibility for management shifted in the north tubewell. The first two committees had been created by people volunteering to take up the leadership. The farmers had not been happy with the performance of the second committee. None of the people living in the village had been willing to take up the leadership role when the third committee had to be formed. According to Shyam Bahadur, he had just come back from the army and learnt that a meeting was taking place in the village. The moment he entered, somebody suggested that he take over as chair person of the tubewell. This was immediately endorsed by the whole village. Even when he had tried to explain that he knew nothing about deep tubewells he was still chosen by the group. They assured him that he would learn. The people had suspected that some funds were being mishandled by the previous committee and no one was ready to take up the challenge of dealing with these older committee members.

One of the first challenges to the deep tubewell committee was to ensure that all farmers paid the correct fee for the demand charge. Many people tried to register smaller areas under the deep tubewell after handover. This reduced the amount of money that each committee could gather to pay for the demand charge. How a committee was able to enforce and implement this rule depends on the integrity of the committee too. For instance, the new committee that took over the north tubewell in 2003 discovered that some area under the tubewell had not been accounted for. The former committee collected a demand charge from only 50 hectares of land. After investigation into the matter, the new committee was able to collect money from 64 hectares. Each deep tubewell committee fixes the rate for demand charge for their individual tubewell depending on the rate set by the electricity authority. They revise the rates whenever there is a rise in the electricity tariff. The committee then notifies the fellow villagers of the revision in village meetings.

The operating cost in 2004[11] was NRs. 170 and NRs. 240 per hour for the south and the north deep tubewell respectively. Each deep tubewell also paid around NRs. 2000 per month for the demand charge. The demand charge was fixed at NRs. 450 per bigha for the south tubewell and NRs. 400 per bigha for the north tubewell. The south tubewell therefore gathered NRs. 23400 rupees as demand charge and the other tubewell gathered Rs. 37600. The total amount was paid off to the Nepal Electricity Authority (NEA). The committees do not face many problems where the collection of the operating cost is concerned. This is because the farmers are 'more willing to pay for something that they use' (deep tubewell chairman south tubewell).

Farmers demanded that the committee open new outlets in the south tubewell. When the BLGWIP was in effect, the farmers did not ask for it. The project allowed an outlet per every fifty metres only. There are more claims by the farmers against the committee, and every farmer wants his or her own outlet. This problem has come up according to the chairman, 'with more democracy' and also because they have now to pay for water. The farmers whose plots are not directly connected to an outlet are the ones who have been demanding this.

Another issue for the committees was to oversee a pump operator. The pump operator is mostly a poor farmer who is also in the village most of the time. The pump operators before handover were regular staff of the project. They were paid a salary by the project. Each deep tubewell has its own way of paying the pump operator. In Supauli, both the committees pay the pump operator a salary of NRs. 10000 per year, which is a nominal sum of money. A pump operator of the project used to draw that amount of money in two months. There are rooms for the operator to stay in the pump house. However they live in their own houses in the village. This village has developed its own way of giving incentives to the pump operator. The pump operator in the north tubewell gets an additional bonus of NRs. 5 for every hour that he operates the pump.

The executive committees of the deep tubewells meet when they feel necessary. There is no fixed date for the meeting in Supauli[12], and there is no fixed venue where the meetings are held. It is sometimes held in the house of any committee member or even in public places[13]. The general body meeting of all the water users is held once a year, usually in the month of May. The committee has also made it a rule that the person who goes to pay the electricity bills is given a travel budget of NRs. 100 per month. They also keep a minimum amount of money for cost of paperwork, for stationery and letter pads of the committee.

The rules for both the deep tubewells are passed in their specific general body meetings. They are documented and form a set of formal rules made by the committee (See Box 4.1). However, they are not always applied as they are written down. For example, it is emphasised that all farmers have to pay for the operating cost at the time they request the water through the pump operator. How a

BOX 4.1: Rules set by the deep tubewell committees

After taking over the management of the tubewells, the committees in both the tubewells devised certain rules and regulations: rules related to water distribution and payments. One of the first rules that they made was setting up a control for the minimum time farmers could request for water. This was done because first of all, it was not convenient to open the pumps for very short periods of time. In addition, the discharge from the tubewells was high: 435 m^3/hour in north and 415 m^3/hour in south tubewell and the carrying capacity of the canals was small. This was because the conveyance system of the deep tubewells had been constructed over the existing kulo. The committee in the south tubewell set up a rule that any person who wants to irrigate, has to at least demand water for a minimum time of five minutes. The committee in the north tubewell set the time at twenty minutes. The north tubewell managers encourage the farmers to come together in a group and request for water.

The other rules are related to payment. All farmers are required to pay the demand charge by the end of April each year. The payment for running charge was to be made at the time the farmers demanded water. When the farmer decided to irrigate for more time then initially demanded, they were allowed to pay the remaining amount at a later time. A continuous time of one hour was allowed to those farmers who wanted to irrigate 0.68 hectare (one bigha).

The farmers are not allowed to grow anything on the shoulders of the canals. However, many of them are growing lentils or fodder grass. The committee decided they could not control this. A rule was formulated in order to gather cash from this activity. Any farmer who grew lentils or grass on the shoulders was expected to pay NRs. 2.50 per metre of canal length, for the length of canal that they used.

Source, Field work 2004

person is made to pay depends on the pump operator. If the pump operator is willing and has good relations with the person in question it is found that he took the money at a later date. Each committee checks to see if all the money has been collected from the farmers. The committee members audit the records kept by the pump operator. The pump operator has to clear the accounts with the treasurer. The accounts are discussed and checked by all executive committee members. But this depends on the accountability of the committee members, as the incident in the north tubewell demonstrates (see Box 4.2).

BOX: 4.2 Obstacles in Building the deep tubewell committee

When the new committee took up the responsibility of managing the deep tubewell, it was found that the former pump operator had not cleared the account for a sum of NRs. 12,000. The pump operator is a poor Tharu farmer who has been working in different capacities as the kulo chairman or pump operator. The new committee decided to charge this amount of money to the treasurer of the previous committee, because it is also the duty of the treasurer to oversee the accounts sincerely. However the same person who acted as the treasurer in the old committee was also in the same post in the new committee. This led to bad feelings between the new chairman and the treasurer. Any person who wants water at a certain time has to pay at that instance for the time that he or she demands water. If the irrigation period goes beyond the requested time, they have to pay again at a later date. Account keeping depends a lot on the credibility of both the treasurer and the pump operator. The pump operator at that time was not able to keep good records because of inadequate account keeping and writing skills. It was not clear who misused the money though.

The new committee opened an account of the water users' group at a local bank. It introduced a system of both vouchers and log book to keep account of the demand for water and the money paid by the farmers. Previously, the committee used to keep the accounts only in the log book. This made it difficult to keep track of the real income. There was a rift in the committee, because the treasurer threatened to quit because he was made to pay the amount that had been mishandled. He installed a shallow tubewell and declared that he would not pay the demand charge and that he would take the shallow groundwater through the deep tubewell canal. The chairman warned him that he would have to pay for both the demand charge and the privilege of transporting shallow groundwater. Two such incidents of shallow tubewells had come up in this village. In the other deep tubewell area, the farmer had been warned to buy his own pipe for transporting the water. The treasurer has been warned by the committee that he would have the deep tubewell canal outlet closed off.

Field work, 2004

The details show that setting up control over groundwater is directly related to the way the actors strategise and find out ways to develop a mechanism for managing the deep tubewells. The committees have not been able to institutionalise ways to gather funds for repair and maintenance. The farmers in this village pay for two different charges in irrigation: the deep tubewell demand charge as well as NRs. 1000 per year per 0.68 hectare for the kulo. In addition the farmers have to contribute labour for Demanding more money on a regular basis for maintenance was not a favourable option. The two tubewells belong to the oldest generation tubewells that were installed by the project. Repair and maintenance is a concern for the committee. This involves changing of minor parts, changing larger and more expensive parts like the transformer coils and also tubewell washing. The spare parts are expensive and costs range from hundreds to thousands of rupees. The project had made arrangements to support the tubewells for repair and maintenance for two years following handover.

In order to execute this, the project had established a co-ordinating committee. This committee encompassed all the tubewells in the whole of Rupandehi district. An executive committee and four regional committees were established. The central co-ordinating committee comprised of all deep tubewells as members of the committee. The regional committees were established with geographic significance. The main function of this committee was to liase between the project and the deep tubewells. A mechanic was appointed for each regional committee. These mechanics were given the responsibility for overseeing the maintenance of the deep tubewells. In order to do this, the mechanics had to inspect each tubewell every month. Each deep tubewell is required to pay a sum of NRs. 1000 to the Regional Co-ordinating office. This payment is made to avail itself of a mechanic. After every visit, each individual committee of the deep tubewell has to keep a record of his visits in writing. If he fails to appear, they try not to pay him. The deep tubewell committees coordinate with the regional committees for obtaining spare parts for the deep tubewells. Any tubewell that needed washing had to apply to the Regional Committee. This committee demanded that the five or six tubewells made one application together at a time. Tubewell cleaning requires a mechanic and ten labourers. It also requires the materials for cleaning the tubewell. For some years the project provided the mechanic and the material. From 2004, the committees had to pay for the cost of the material and mechanic, which involves his daily wage as well as his transportation costs. When the project gave the support it was not difficult. The support has been taken away and the farmers have been trying to find out new avenues for this. Even though BLGWIP had withdrawn from providing the finance for washing, the project still supplies them with the mechanic if they want him. It also buys the material for them if they pay for it.

The committees have found ways of reducing costs of tubewell washing. The committees in Supauli arranged their own cleaning at half the cost that the project estimated. They performed cleaning by mobilising all farmers who irrigated from the tubewell. The committees also tried to muster funds for the deep tubewells from different sources. One of the committees, decided to gather money by selling the wood from the trees around the premises of the pump house. With this money, they repaired the gate of the pump house as well as paid for tubewell washing.

Deep Groundwater and Shallow Groundwater Use in Tikuligarh Gaon

The second study in this chapter is based in Tikuligarh gaon, where the farmers use only groundwater for irrigation. The tubewell is here 'running a thin line' between continuation and closure. At the end of the field work, the deep tubewell was on the verge of closing down. The village was also not able to retain its rights to the kulo. In this sub-case study, I examine the struggles of the farmers in Tikuligarh gaon in the process of incorporating deep tubewell irrigation and their struggles around different sources of water. The story of Tikuligarh gaon is very different from that of Supauli, even though they lie next to each other. While farmers in Supauli are trying to gain control over the different sources of water, the farmers in this gaon are losing their control over the deep tubewell. A few actors have been struggling to retain the deep tubewell. However, it has become a very difficult task for them: social relations in the village and the relations in surface irrigation have spilled over in the management of the deep tubewell.

Losing rights to the kulo

Tikuligarh gaon, like Supauli, is also one of the oldest mauja of the Sorha Mauja irrigation system. A large part of the area consists of settlements characterised by very small holdings. The other portion of it is made up of up of settlements of farmers with relatively larger land holdings. Forty-nine percent of the 352 families living in this ward are very small farmers and own land of less then 0.34 hectares of land. Twenty percent own from 0.34 to 0.68 hectares, nineteen percent own from between 0.68 and 1.4 hectares, nine percent own 1.4 to 3.4 hectares and one percent own more than 3.4 hectares. The majority of the farmers are owner cultivators. Most of the smaller holdings are owned by hill migrants who also have very small businesses like alcohol shops or teashops. A majority of them also work outside the villages. The other portion is a group of relatively better off hill migrants from different castes and ethnic groups. There is also a large community of ex-army men or policemen.

The way settlement developed along the kulo here has had a bearing on the social relations in the village and on irrigation. Most of the migrants in this kulo area settled down in its head end part. A distinct grouping of farmers developed with the older irrigators located at the tail end and new irrigators located at the head end. Most of the migrants, according to an old kulo chairman, 'could not clean the kulos like the Tharus did'. The migrants used the short-handled hoes *(kodalo)*, that they had brought when they had migrated from the hills. These hoes were not suitable for digging canals in the Terai, where farmers use long-handled ones. The tailenders on the other hand, claim that head end farmers did not give enough water to them. Some migrants admit that it would have been much better if the Tharus had been living in the upper reaches. They are more skilled in the art of construction and maintenance of the kulos. The canals had been constructed with a certain alignment so as to irrigate the present tail end villages. This was spoilt as migrants started tampering with the water flow in the upstream portion of the canal.

A large part of the tail end belonged to the Chaudhary. He filed a complaint against the kulo chairman at the office of the zonal commissioner, claiming that he was not getting a fair share of water from the kulo. The management of the kulos had already shifted well into the hands of the hill migrants. The kulo management claimed that the landlord was cheating by not contributing the real amount of labour required for canal cleaning. According to the kulo chairman who was involved in the incident, he was summoned at the office of the Zonal Commissioner and was asked to give explanations. The Commissioner was not aware of the rules and regulations regarding irrigation kulos. When the rules were explained, the ruling was made in favour of the kulo committee rather than the landlord, because according to the chairman, 'the administration did not find it appropriate to interfere in matters related to the kulo system when they found out that the kulos are managed on certain principles.' He added, 'governing the kulo is different from the governing the state.' Kulo management faced problems from other farmers also. Farmers were refusing to participate in the canal cleaning works. They tried to justify their behaviour on the basis of the behaviour of the landlord. They claimed to take part only on the condition that the landlord was forced to participate in contributing the full amount of labour that was due for the area that he irrigated. As many farmers did not participate, the kulo management had to pay the fines to the main irrigation system. It was finding it very difficult to cope with the financial burden.

The project started drilling the deep tubewell in 1978. The tubewell started operation in the middle of 1983. Many farmers in this kulo system let the project line the existing surface canals. After having permanent lining on their portions of the canals, these farmers refused to take part in the canal cleaning operations. The kulo committee declared that those farmers who did not take part in canal cleaning would not get water from the kulo. This led to a division amongst the villagers. There were two groups: those in favour of kulahi and those against kulahi. After getting free groundwater they stopped going for kulahi altogether. The leadership could no longer control the surface source. They discontinued their rights to Sorha Mauja.

Also an incident that took place in 1966 still generated repercussions. It was basically a case between two prominent figures in this village: One of them was a migrant Anand Thapa, who came to live in the village in 1961. The other was the Chaudhary. Thapa claimed that he had resettled 55 hectares of land in the area for different ethnic groups and castes, when there were no government-sponsored resettlement programs in this particular area. It is not clear what political group he then belonged to, as most political parties were banned in that period. The general public just knew him to be close to the government. After 1990, he joined the democratic Nepali Congress party. This person had been active in helping hill migrants to settle in the area in that period (personal communication, February 2004). Most of the migrants in this VDC settled in after buying land from the Tharus. However, there are small patches of land where people cut the forest and settled down[14]. These areas were not irrigated by the kulo. However large parts of the present day VDC were controlled by the Chaudhary. The migrants, along with a group of people from another village outside the present VDC, were involved in intimidating the landlord. This was in the period after the declaration of land reforms and also a period when all political parties were banned. Most of those involved had leftist leanings and wanted to intimidate the landlord. The incident took place when the Chaudhary was holding the meeting of the Sorha Mauja in his yard. One of the persons from the group had attacked the Chaudhary physically. There was a scuffle which was then mediated by the family members and another migrant who lived in the same village. 'There were gun shots from the house'. Later he filed a case against forty-six people. They were arrested and later released on bail. This incident had already created rifts in local society.

Problems around groundwater

The deep tubewell was the only source of water after the farmers abandoned their surface water rights. It was also the busiest tubewell in the whole area. The water users' committee sometimes also provided water to Durganagar which did not have a deep tubewell. They transported the water through the jharan canals. After handover, the farmers reduced the use of deep groundwater like in other tubewells.

Some farmers had started installing shallow tubewells before the deep tubewell was handed over while others installed them later. Some farmers obtained shallow tubewells through the group subsidy provided by the bank. A single farmer later paid off the interest from the group and turned it into individually owned shallow tubewells. The large landowners were not willing to pay the full demand charge for the deep tubewells. All farmers who installed shallow tubewells claimed that only a small portion of their land was actually irrigated from the deep tubewell. This reduced the amount of demand charge collected by the committee. It therefore raised the rate of demand charge for the rest of the farmers who do not own shallow tubewells. Other farmers who bought little water from those farmer who owned pumpsets, also started to lay similar claims. The deep tubewell committees were not in a position to put a control on this behaviour because one former chairman had installed a shallow tubewell himself.

The tubewell was designed to irrigate 187 hectares. In 2004, the deep tubewell committee collected demand charge from only 35 hectares of land. The deep tubewell committee hiked the demand charge to a rate of NRs. 1000 per 0.68 hectares per year. A reduction in area from 187 hectares to 35 hectares has therefore reduced the income from demand charge by 81 percent. There are eleven shallow tubewells in the command area of this deep tubewell. Thirty-five hectares of land have shallow tubewell facilities. This does not include the area of farmers who buy the water from these pump sets.

PICTURE 4.1 A pumpset installed next to the deep tubewell distribution system

The process of establishment of a committee to manage the tubewell has not been easy. No committee is willing to stay beyond its term. A fifth committee was in charge of the deep tubewell in

2004. The present chairman was also the first chairman to take over the deep tubewell after handover. However, neither the first nor the other committees that followed could make the shallow tubewell users pay for the total demand charge. Farmers had installed shallow tubewells and were transporting water through the canals of the deep tubewell system and also refusing to pay the total demand charge. People gave different justifications for not paying the demand charge. For example, a migrant farmer with a considerable size of land holding of around 3.5 hectares declared that he would not pay the demand charge because he was no longer using the deep tubewell. He has a shallow tubewell and irrigates from that. He gives the justification that, if the landlord can stay in the village without paying the demand charge, why should he pay for it when he too is not using the deep tubewell?

The responsibility for managing the tubewell is now revolving back to the same persons who had been working in different capacities in the previous committees, even if they are not willing to do so. The deep tubewell has to be maintained because a large section of the population still relies on it.

PICTURE 4.2 Farmers getting ready to clean the deep tubewell in Tikuligarh Gaon

Efforts at managing deep tubewells

The problem of the shallow tubewells' spread was so great that the present committee wrote formal letters to all those who owned a shallow tubewell to come and pay the complete demand charge. No one responded to it. When the BLGWIP was in effect, the project presence was so strong that there was some control over shallow tubewell installation in the project area. The shallow tubewells are a personal choice of the farmers and there are no legal rules preventing their installation.

The deep tubewell committee made certain rules and regulations regarding the distribution and allocation of water. The discharge from this tubewell is highest in the area and is at the rate of 545 m^3/hour or 0.151 m^3/sec. That means that, even in a short time, the volume of water obtained from the deep tubewell is high. The committee set up several rules after handover. The minimum time a person could demand water was set at five minutes. However, this was found to aggravate the motor. They later raised it to fifteen minutes. Two farmers sometimes requested water together for fifteen minutes. This sharing was possible only between those farmers who paid the demand charge. After every fifteen minutes, the pump operator closed the tubewells, even when others waited in line for water. This too was not practical for the motor. Each person was then made to wait five extra minutes, so that the person who started to irrigate first, got the total volume. This was based on the time that it would take the canal to drain all the water. As the discharge from the deep tubewells was high, small plots could be irrigated very fast. Farmers with large landholdings benefited from this volume. They pay the demand charge for only a portion of the land that they owned, but they irrigated at different stretches of time and obtain the volume that they require. Some farmers paid the cost of transporting the shallow groundwater through the deep tubewell canal at the rate of NRs. 1 per metre.

Deep Groundwater and Jharan Use in Durganagar

Deep groundwater irrigation in Durganagar started nearly seventeen years after it was practiced in Tikuligarh and Supauli. A deep tubewell was installed in Durganagar only in 1999. The farmers in this village used the jharan and shallow tubewells for dry season and winter irrigation before a deep tubewell was installed here. They used to irrigate from shallow tubewells in the eighties. It was logical for the people in this village to wish for a deep tubewell, when they saw their neighbours irrigate with free abundant groundwater. In this case study, I examine the process by means of which different actors in Durganagar pursued to obtain a deep tubewell for irrigation, how they incorporated deep tubewell irrigation into their existing system and how they have worked out ways of preserving the existing source of water.

There were 155 households in Durganagar in 2002. There was only one person in this village, who owned more than 3.4 hectares of land in 2004. Nine farmers owned land in the range of 1.4 to 3.4 hectares. The distribution of those within the ranges of less than 0.07 hectares, 0.07-0.34 hectares and 0.68-1.40 hectares was similar. The proportion of hill migrants to Tharus is almost equal in this village and there are five families of Lodhs. Majority of the farmers are owner cultivators, while some hill migrants made sharecropping arrangements for vegetable cultivation.

Harnessing the jharan

Before the shallow tubewells were installed, the only source of water for this village was the sub-surface spring water that comes up from a patch of land north of Tikuligarh village. Historically, the farmers in Durganagar had been harnessing this jharan and irrigating from it. In addition to this, they also re-used the drains from Tikuligarh village. However, this drain water was available only when Tikuligarh village irrigated from the Tinau River. After Tikuligarh village stopped using surface irrigation, this source of water was no longer available for the village. The only source of water was the other jharan, which the farmers refer to as the jharan 'without a source' or with an invisible source.

Seventy-six hectares of land in Durganagar have access to this spring source. The jharan typically starts to flow in the village in the second week of June, although it depends on the time of start of the monsoon. In 2002, the water was already flowing from the 10th of June, before the farmers had finished clearing the whole canal. They blocked the water and started distributing it only from 17th June. The jharan normally starts flowing to mid-kulo height by the middle of June. The water is distributed by turns when the flow is low. It is then left open as soon as the monsoons start.

Maintaining the jharan is important to Durganagar. The farmers

FIGURE 4.2 Jharan and DTW pipe distribution (underground) system in Durganagar

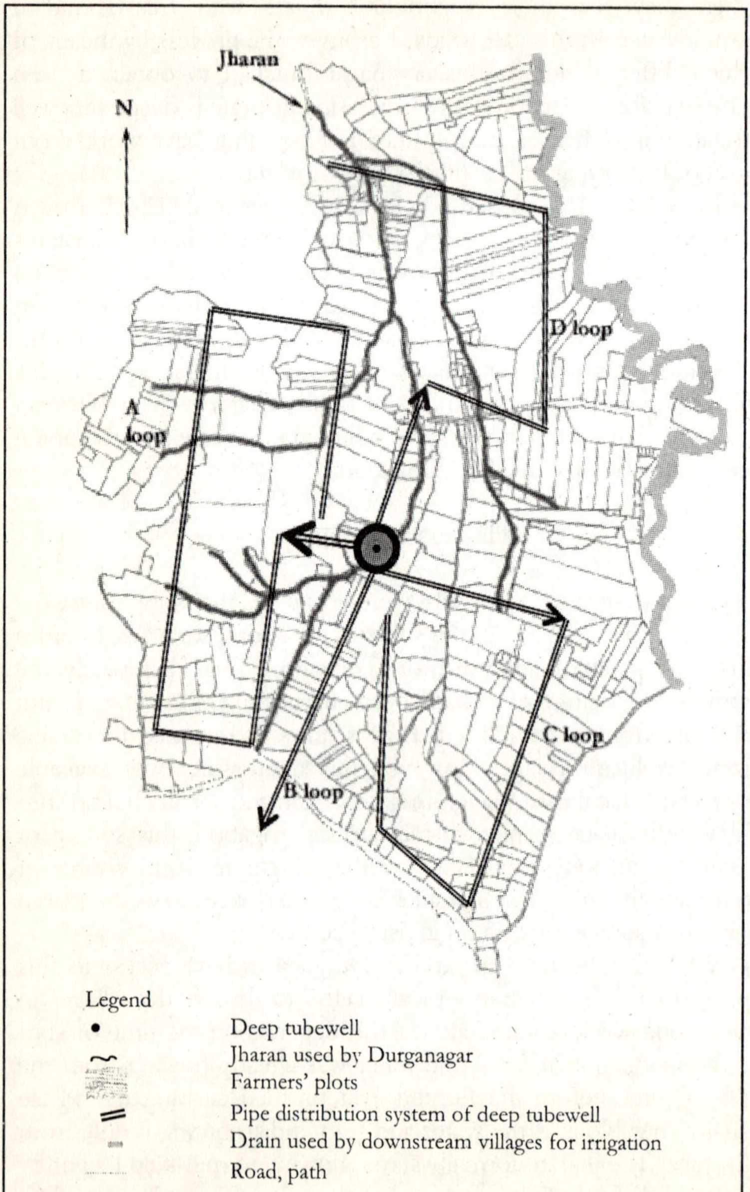

Source: Field Survey, 2002/04

prefer the quality of the water and it is also a cheaper medium of irrigation. In addition, the flood water has to be drained off every monsoon. The jharan kulo is dug deep so as to accommodate the flood waters. It takes the village four days to clean the whole length of the jharan kulo. This is the length that they clean from the source to the village. The work at the source is nothing in comparison to the amount of work the farmers using the kulos perform at the Tinau headworks. The only difficulty they face is when they have to clean the canal which passes through settlements. They have to take considerable care to carry out their work with discipline and tidiness so as not to disturb their neighbours. This is particularly important when the canal passes in front of houses. The villages through which the jharan canal passes do not take water from this source.

The village has its own system of mobilising the cleaning operations for the jharan kulos. The person in charge is called the mukhtiyar, like the chairmen of the kulos. He took up the charge of the jharan after his father, who had done it after he migrated back from Burma. The jharan in-charge is equally active in mobilising labour for the jharan, as well as in issues around the deep tubewell. He has also been working as member of the deep tubewell committee for two terms. His experience in managing the jharan is an asset for deep tubewell management as he is well aware of the pattern of landholdings across the village. This knowledge comes in useful particularly when the deep tubewell committee re-checks the accounts of the deep tubewell, to see if there has been any cheating on the payment of demand charge.

Each year, the jharan in-charge fixes a date for the cleaning operations. This message is conveyed to the whole village through a messenger. All farmers are expected to attend. There is a fine for the absentees. Labour obligations are tied to the area of land under the jharan. All farmers with land of 0.68 hectares are required to attend the cleaning operations for one day; those who have 1.36 hectares have to attend for two days and so on. The jharan in-charge maintains a register in which he notes down the number of hours that each person has to contribute. He takes the help of one more assistant for managing the task. The length of kulo that each farmer has to clean is measured by the leader by means of a stick. According to the jharan in-charge, some farmers tried to avoid the cleaning activity after they got the deep tubewell in the village. The leaders are trying to maintain the cleaning operations not only for irrigation but also as prevention against floods. Most farmers still participate out of the fear of the flood waters. In 1993, the whole village was flooded even after they had cleaned the jharan.

We could observe that there was a close collaboration between the VDC Chairman, who also lives in the same village and the jharan in-charge. The VDC chairman came to check the cleaning operation on all four days. In his role as the leader of the village, and the VDC, he was concerned for both irrigation and flood protection.

Procuring water from other deep tubewells

Before the handing over of the deep tubewells, the farmers in Durganagar sometimes requested the pump operator of the deep tubewell in Tikuligarh gaon to give them groundwater. Not all the farmers in Durganagar had shallow tubewells; and those who did, had to spend their own money for operating them, while their neighbours irrigated at no cost. There is no direct canal connecting that deep tubewell with this village. The farmers used the canals of the jharans and kulos to transport the deep groundwater when they wanted it. When and how they got water from the deep tubewell in Tikuligarh gaon depended on the relationship the people from this village had with the operators and managers of the tubewell in the upstream village. Some farmers complained that it was difficult when a particular person was in charge of the tubewell. According to them, he was from the Nepali Congress party, while most of the 'active people' in Durganagar who took the initiative to request water were not. The people of this village, got a written permission from the project to put pressure on him to open the deep tubewells for them whenever he hesitated to open the pump for them. This shows that the people of this village were very active in networking with project authorities and across the VDC and applying pressure to get water. There was also another cause for discontent of this village with the 'water managers' in the upstream village. The same tubewell manager had also been in charge of the kulo when the gaon had discontinued its surface water rights. Durganagar did not irrigate from the Sorha. But when the gaon stopped irrigating from the Sorha, one of the water supplies to this village was also cut off.

The northern part of this village relied on jharan from the Chaudhary's large landholding. The exit of the jharan from Chaudhary's plot entered Durganagar directly. This supply of water arrived earlier in the village than the water from their main supply: subsurface jharan. The Sorha Mauja draws its water from the Tinau, and these villages irrigated earlier than Durganagar. Hence the drains from these villages were an asset to the farmers in this village.

This village maintained good relations with another chairman of the deep tubewell and got water when they requested. He was a party worker of the Rashtriya Prajantra[15] party, but according to the leaders of this village, 'they could not afford to make him their enemy as the vote from the Magar and Gurung community depended on him'. This was important, because the major political parties in the election are the UML and the Nepali Congress. There was a competition between the workers of these parties in trying to woo the other minor parties on their side, so as to get a majority in the elections[16].

Efforts for a deep tubewell

Even though the farmers of Durganagar had always wanted a deep tubewell, they were not able to put forth their demand formally until 1988. This was possible only in the third phase of the project. Four people in Durganagar are known to have been very active in 'bringing' the deep tubewell to the village. This group was very active in maintaining contact with tne project authorities in order to obtain a deep tubewell. Besides irrigation, other facilities are also tied along with the project package. These include gravelled roads and three-phase electric connections. The village was connected with village roads, but it was still more 'remote' than other neighbouring villages. The people who took the responsibility for pursuing for the deep tubewell, mobilised all the people in the whole village and put forth their demand at the project office. It took several years for the project to be effective because of the changes in the political situation in the country. Implementation processes were slowed down because the whole system of government was undergoing restructuring after the democratic movement in 1990.

The leaders in this village were in constant contact with the project. The first 'official meeting' between the project and the farmers was held in March 1996. The informal meetings and negotiations had been carried out since 1988. Initially, many people were sceptical about the deep tubewell, as they had observed the struggles of the farmers in the neighbouring village. Later they were convinced about the benefits of the tubewell. One of the first activities for a project with a 'demand-based participatory approach' is to send social mobilisers to convince the people of the benefit of the incoming project. The institutional development unit of the project was the main division responsible for this part. However, the local leaders of this village were more involved in convincing the people about the benefits of the deep tubewell and the new water distribution system. The farmers were aware that there were to be changes in the design of the tubewell: the distribution system was to be an underground UPVC system. They were told that the discharge from the pipes would be lower than from the open flow system and that the tubewell would be more economical than older tubewells.

The project did not really need the 'social mobilisers' for this village. This was stated by one officer who had been involved in the social mobilisation unit of the project. According to him, Durganagar was the easiest place for implementation of all the villages. The village really needed groundwater. The upland parts of the village could not access the jharna. At the same time, the whole village needed water for winter and dry season irrigation. Many farmers had been using shallow tubewells during these times and were a bit sceptical about the cost of the deep tubewells. The leaders and the social mobilisers from the project convinced them that a deep tubewell would be cheaper than shallow tubewells.

The participatory concept of the project required the farmers to contribute to the construction of the deep tubewell irrigation system. It also proposed to involve them in the design. Involvement in design was related to the participation of the farmers in the layout of the loops. The water users had to contribute 25 percent of the cost for layout of the distribution system. The village gathered NRs. 36000 in bighatti (per bigha of land owned). Besides this, they contributed labour to make up the 25 percent of the total cost. The farmers planned the layout of the four loops in such a way that it covered the whole village. In the process of laying down the pipes, they also made sure that the loops did not cross over the canal carrying the jharan water, as it would obstruct smooth cleaning of these canals and would trap the debris that gets carried in times of flood. There are four loops in the distribution system. Three loops were enough for this village. They also convinced the farmers in the neighbouring VDC to share one loop.

The tubewell started operation from 1999. Most farmers in Durganagar had either sold their pumpsets or stopped the shallow tubewells by 2001. A few farmers kept their pumpsets. They used the shallow tubewells in emergency, when there was an electric failure. The traditional vegetable farmers, the Lodhs did not sell their shallow tubewells. They did not want to take the risk of electric failure in crucial irrigation periods. They tried their best to use the deep tubewell overall.

The people of Durganagar paid the lowest amount of money as demand charge in comparison to other tubewells because of a lower capacity motor of 37.5 kW. The rate in 2004 was NRs. 100 per 0.68 hectares. This was one-tenth of what the people in Tikuligarh village paid for their tubewell in the same year. The running cost was NRs. 160 per hour. When all the loops were in use for irrigation, the rate per hour reduced to a lower value and cost the farmers NRs. 40 per hour.

Evolution of the committee for managing groundwater

Four people were very active in taking up leadership roles to pursue for a deep tubewell for this village. The first committee comprised of seven members in its executive posts. It included the four active farmers and three additional members. There was also a pump operator. A second executive committee consisting of eleven members was formed in 2004. Five of the members in the second committee were the same persons who had served in the first committee.

The idea and the decision to increase the number of people in the executive post in the second committee came from the members of the first committee. According to them, it was done, 'so as to get a representation across all castes, ethnic lines and gender'. The second committee included the loop leaders, people from different castes, ethnic groups and also a woman. The same pump operator was hired again. Most of the former members were deeply involved in party politics. They needed time to prepare for the elections. The former VDC leader opted to stay in the capacity of a member only. One of the reasons for the politicians not to leave the deep tubewell 'connection' is because of the success of the tubewell.

This is the one tubewell that became 'famous' for successful operation amongst all the tubewells that had been installed in the district. This was one of the biggest changes I experienced from 2002 to 2003. People from different agencies were visiting the tubewell in order to see how it was operated. The number of visitors had increased so much, that the committee had arranged chairs for the visitors and a visitor's book. Any visitor to the pump could sign his or her name into the register and gives comments or suggestions for further improvement in management. A visitor from any bilateral organization or international non-governmental organization was charged NRs. 500 for the visit in 2004. The helpers got a return receipt for that[17]. A rule had been set, that half of this money was put in the account of the committee, while the rest of the money was given to those persons who are involved in briefing the visitor about the deep tubewell and taking them around the fields. Any other persons like researchers or consultants who came on their own and wished to see the pump in operation (in the times that it is not in operation) 'could' contribute cash. The committee had not fixed a definite amount for such people. It was left to the decision of the researcher or the consultant. This money was put in the deep tubewell fund. The committee made a rule not to charge any money in case of student visitors. This rule was set up in order to compensate for the time that the farmers or pump operator take off from their busy schedules so as to provide time to visitors.

A general body meeting of all farmers is called by the committee, whenever they have to make crucial decisions that require the consent of the majority of the farmers. The first 'formal' meeting between the farmers and the project took place in March 1996. This was done to formally inform the farmers about the incoming tubewell. There have been seventeen meetings in the village regarding the tubewell. Five out of these have been general body meetings of the members of the deep tubewell. The other was those between the executive committee. These meetings involved decisions regarding repair or maintenance. Even though this is a relatively new tubewell, the farmers were experiencing some difficulty in operation. The coordination between different loops was a problem.

There was a difference in water demand between loops. The farmers in one loop area (loop A) did not need much irrigation water because the area fell on low lying land, where the soil moisture was relatively high throughout the year. The demand for water was highest in loop D, as most farmers in this loop grew vegetables. After the deep tubewell had been installed the area under cultivation had also increased in this part of the village. Another loop area that demanded irrigation water regularly was from the farmers in loop C. The farmers and committee members felt that they needed a modification in the design and layout of the loops. In 2004, they made a decision to join the Loops C and D. Some of the pipes had started leaking and needed to be changed. The committee decided to carry out the operation of changing the design of the distribution system. It was agreed upon by consensus by the farmers of the village in the meeting held on 13 December 2003.

The committee is still in the process of learning from experience and making rules and regulations as they come up. Rules for payment of the demand charge and operating cost were set up from the start. Each farmer was required to pay the operating cost by the 22nd of the month in the Nepali calendar. Late payment entailed an additional fine to the farmers. This was done so as to make up for the fine imposed by the NEA for late payment of electricity tariff. The demand charge collection was done once a year just before the peak irrigation season, in the month of May.

There has been one incident of water stealing in this tubewell 2003. A farmer who had not paid the demand charge asked his neighbour to request two hours of irrigation from the deep tubewell. The pump operator had conceded to the demand. He later got suspicious because he was aware that it took around four hours to irrigate one hectare of land in the month of May. The farmer who had come up with the request had only 0.17 hectares of land and it was obvious that he did not need two hours of irrigation. The pump operator found that the farmer, who had made the request through the friend, had not paid the demand charge. The committee warned the neighbour for letting his friend irrigate illegally. The committee made a decision that they would not let him irrigate again unless he had cleared all his dues for the past year. The executive decided to punish the pump operator as well for not being vigilant while carrying out his duty. They decided to cut off a part of the salary of the pump operator if he was unable to collect the money from the farmer before the next paddy season. The pump operator of this tubewell is the highest paid pump operator in the whole area. He was paid twice the amount that his counterparts in other deep tubewells were paid.

He was also the busiest because this tubewell was the tubewell with the highest use hours amongst all the other tubewells in the area. The pump operator was waiting for the farmer to come and demand water in the summer of 2004. He had decided that he would not let the farmer irrigate unless he paid the dues for the past year as well for the coming year.

The records of the tubewell show that substantial saving. The income expenditure statement of the deep tubewell is shown in Table 4.1. The saving at the end of 2003 was nearly NRs. 59000. This is a substantial amount for a tubewell that has been in operation for less than four years. Some other tubewells in the VDC did not generate money for savings, and their only finance was money put in by the VDC (more on this later) that was kept as fixed deposit. It can be seen that the tubewells draws large amount of money from operating cost which makes up for the portion of the demand charge also.

TABLE 4.1: Income Expenditure Statement of Durganagar Deep tubewell (April 2002 - December 2003)

INCOME		NRs.
Bank Balance		68305
Collected from different individuals		7082
Money obtained from VDC		23000
Collected from Bighatti		11000
Interest from Bank		5238
From the four loops (the receipts)		113248
	Total	227865
EXPENDITURE: till December 2003		NRs.
Electricity		118780
Pump operator salary		30000
Miscellaneous		1190
Purchase of chairs		1610
Payment to regional committee for 3 years		2500
Repairs		12529
Printing of vouchers (twice)		1830
Wiring		560
	Total	168999
Amount remaining		58866

Source: Water users group of the tubewell, 2004.
Exchange rate for 2003: 1US$=NRs.76.14

The total amount the tubewell paid for electricity was NRs.118780 for the period April 2002 to December 2003. This included payment for both electricity tariffs: demand charge as well as running charge. The money collected from the four loops was NRs. 113248. This is the money collected only for operating cost. This itself paid off a large percentage of the electricity fee for the tubewell. The money collected for demand charge is the one under 'money collected from bighatti'. This was NRs. 11000.

A discussion of the individuals who play a significant role in managing of the deep tubewell and the jharan in this village is also a discussion of the main actors in the VDC who have also been making significant changes in the VDC. This is because they worked both as residents of this village and in a network of people who were deeply involved in decision making for water resources management in the VDC also. One of them is Ram Prasad. He was the elected ward member in this particular village till 2002. A college graduate, Ram Prasad, performed the role of a 'public relations officer'. An intelligent, soft-spoken and well informed person, he had all the characteristics of a successful 'networker'. He was the main person who led the negotiations with the projects and officials in different sectors for the whole VDC. He was involved in the process of dealing with the project right from the beginning along with three other persons to bring in the tubewell to the village. He had the support of the others. Another person also active in the village social activities was Hari Bahadur. Both Ram and Hari, came from the western mountains to settle in this village. The other two were: the jharan in-charge: Krishna, and Pandey, the VDC chairman. Both of them came from Burma to live in the village on the same day. They came with their parents and relatives. Similar political ideology holds all these people together.

Throughout the fieldwork, it was obvious that the VDC chairman was well liked by most people, irrespective of which political party he belonged to. The village and VDC wanted a change from the old leadership and this group represented young and enthusiastic people who the people 'thought could do something'. Only those people who were genuinely involved in 'party politics' felt it necessary to criticize him. He came to power immediately in the first general elections in 1991 and won a second term again in the elections held in 1995. This is a large VDC and he definitely could not work alone. He had a whole network of people who worked for him. During conversations slight differences amongst these people became manifest. But on the whole it is very clear that they were working towards building his image for the next elections. During the time of the first part of the field work the local level politicians were building their efforts for another election was supposed to take place in 2002/03.

Krishna was the one who got things done: one who does the actual 'dirty work' in the group. He was more involved in the actual activities within the village. A farmer leader, jharan in- charge and a person contacted by agencies to organize farmer training in the village. With this core group, the chairman got his work done at his village level. But at the same time they were just a part of his network of people who worked throughout the VDC and the district level.

Role of the VDC to support the deep tubewells

Each individual tubewell committee was challenged with the task of managing their respective tubewells because a large part of Tikuligarh VDC needed deep groundwater. However, the effect of handover was also felt collectively by the VDC management. This VDC made its own efforts to provide support for the tubewells.

Each VDC was allocated an annual budget of NRs. 500000 by the government. A VDC is comprised of nine wards. This budget is then allocated in equal amount to the nine wards. There has been a reduction in this budget after the government reduced it for security purposes from 2001/02. The money that reaches the VDC decreased to NRs. 366000. Each ward presently gets only about NRs. 40,000. This is the budget allocated for 'development'. The VDCs were in no position to help the deep tubewells in financial crisis. First of all, expenditure on deep tubewells does not formally qualify as a candidate for the 'development budget'. Second, even if it did, the sum required by the tubewells is too high. Each large deep tubewell has to pay around NRs. 24000 per year for the flat rate of electricity only. The budget of the whole ward is only NRs. 40,000.

In the 1991 democratic elections, the newly elected VDC chairman was under immense pressure to support a large number of tubewells in his jurisdiction. The VDC appointed the ward member and chairman of the deep tubewell in Supauli to formally investigate the actual situation of all deep tubewells inside the

VDC. After this was done, the people who were active around the chairman discussed how they could sustain it. There was a whole network of supporters or friends (but also a politically driven group) that he relied on. They met regularly on an informal and formal basis. They were still active even after the local level government bodies were dismantled in July 2002 and when they were not in power. This networked comprised of aspiring politicians, school teachers and many farmers in the village. Prominent among them were those people mentioned in the case study in Durganagar and Supauli. In addition, there were accomplices in each ward across the VDC. This is a very strong network of people. However, when still an elected chairman, the VDC chairman had to discuss the matters with all the elected ward members, not all of whom were from his own political party. In the year 2000, the leaders decided to open a trust fund to support the twelve deep tubewells in the VDC. They had some monetary resources to draw from.

Privatisation and liberalisation policy of the government after 1990 allowed people to start up their own enterprises. One of the most lucrative businesses that had come up in this area was the gravel industry. There were three stone crushers in the VDC. Besides the crushers, the people in this VDC had invested in businesses that make gabion boxes from wires for river control. The clients were government agencies responsible for river training and road construction works. In addition, the VDC was collecting different types of taxes. One of the important one was the road tax, which was collected on a regular basis. Many trucks and tractors come to this VDC to fetch the gravel and sand from the river. The budget of the VDC had increased to NRs. 7000000 by 2000/01.

Each deep tubewell was allotted NRs. 25000 as an initial fund. Each committee had to keep this a fixed deposit in the Bank. They were allowed to use only the interest. The first amount was given in 2000. In the subsequent years, the VDC gave NRs. 12000 to every tubewell. This is very unique and such VDC support has come up only in this VDC.

Some tubewells irrigate portions of other VDCs also. In that case, these tubewells got money for only the portion of land that was inside this particular VDC. There was an agreement when the fund was established. The tubewells could use the money to cover demand charges only when there was a real crisis. In addition, it was also meant to support the salary of the pump operator. The support obtained from the VDC became most beneficial for the new tubewells like the one in Durganagar rather than the old ones in Supauli and Tikuligarh. However, all the tubewells were still running in this VDC with the support of this fund.

Conclusions

The way farmers in each area have incorporated and adjusted to deep tubewell irrigation is quite specific and localized. The difference was not just between the older tubewells and the new ones. There is a difference between the old tubewells too. Even though the tubewells in Supauli and Tikuligarh belonged to the same generation of tubewells in terms of design and were also subject to similar processes of intervention, the process by means of which the farmers incorporated deep tubewell irrigation is different. The farmers were actively involved in trying to adjust the groundwater technology in the local context. They tried to 'fit in' deep tubewell irrigation into the existing water cycle through different constraints and opportunities at different points in time and in response to different uncertainties. This took place both before and during the initial period of intervention, as well as after the transfer of the tubewells. From the point of view of intervention, groundwater was conceptualized as the major source of water. From the farmer's point of view, it was an extra source of water added on to the existing hydrological system. This is evident from the fact that the farmers in most of the maujas maintained their rights to the surface irrigation. The quality of the surface water is one of the key factors driving the farmers to continue with surface water. Moreover, large parts of these areas are crossed by natural drainage pattern. The farmers have been maintaining the drains and using water for irrigation. Most of the farmers (in maujas that were able to retain their rights to the surface sources) used both the sources of water together right from the start. This sort of attitude ultimately affects the 'expected level of performance' of the tubewells.

The case studies show how social relations have shaped the way irrigation is being carried out in the case study areas. The deep tubewells either added on to or interplayed with the existing forms of social relations in the area. The technology helped to reproduce both competitive as well as cooperative forms of relationships right from the start. Initially, intervention entailed an interaction with the local power structures. The use of power by the landlord to reject a deep tubewell and the acceptance of two deep tubewells by the other group of farmers brought about the first major transformations or differences in the way people made their choices in the source of water for irrigation in all the three sites. Furthermore the conflictive relationship in society in Tikuligarh gaon, and the problematic relationship between the new headenders and the old headenders turned tailenders, was reproduced in kulo irrigation and ultimately in deep tubewell management. The inegalitarian social structure in the same village supported farmers with larger land holdings to install shallow tubewells. This helped to further accentuate the problems of fostering control for people who were responsible for managing the deep tubewells. When the farmers were getting free deep groundwater, the problems in deep tubewells did not come to the forefront.

The role of different actors in shaping the course of irrigation in all the three cases was significant. Transfer of tubewells set off parallel sets of activities with respect to irrigation management. The farmers were equally involved in finding out innovative ways to manage the deep tubewells and at the same time, in trying to gain and maintain control over surface and jharan sources. The strategies that they develop in order to 'keep the deep tubewell running' and the kulos and jharans functioning are very important. In other words, they are involved in both the specific management tasks related to the deep tubewells, kulos and jharans as well as in actions and decisions that supported the deep tubewells at the VDC level. The role of the actors, their authority and the means by which they make decisions, the choices that they make, and the factors driving these choices are very important in shaping the course of irrigation.

The process of transfer and the coming of the new set of deep tubewells coincided with the period when significant political changes were going on in the country and also at the village level. The deep tubewells were handed over to a water users' group but this became almost synonymous with handing them over to the local level politicians. The first sets of managers that took the lead to manage the deep tubewells were also aspiring politicians in the new democratic system that emerged in the country in that period. The deep tubewells were handed over to a so called 'neutral' water users group with practically no formal role for the wards and the

VDCs. Yet the people who took up the responsibility for the deep tubewells were all involved in the local party politics at that time or later. It is usually the same group of people that work in different leadership roles in the villages. The kulo and jharan in-charge, the ward members, the important posts in the deep tubewell committees are all held by the different individuals interchangeably. Besides being involved as chairmen in irrigation related activities the same persons are also members of other co-operatives, livestock committees, and numerous groups that are set up by developmental agencies at the village and VDC levels.

Sharma (2001), in his study on rural water supply in the Terai reports similar findings. He states that politicization is not avoidable and it is not totally undesirable, if the leaders work for the larger benefit of the people. The tubewells were transferred to the water users group. However, in Tikuligarh VDC, the politicians have made a very strong entry into the issues related to governance of water at the local level. It is the only VDC that is directly involved in supporting the deep tubewells at such a large scale. The involvement of the VDC in supporting the deep tubewells is one of the main factors that have led to the continued operation of the older sets of deep tubewells. The political meaning behind the act is also clear. It has reinforced the image of the local leaders and the VDC chairman of a particular political party. The politicians made use of their strong network and took the opportunity to make the decision at a time when the same party was in a majority in the VDC body. They network, and strengthen their political hold at the same time but in the process shape the direction and pattern of irrigation management. The decision to make rules of joint use in Supauli and the entry of the VDC in supporting the tubewells are very important because they have shaped the direction of irrigation in the area. The VDC was able to start up a trust fund for the deep tubewells because of the higher level of income in the VDC. However, other VDCs that have a sufficient level of income have not been able to make decisions in this regard. This can also be said to depend on the level of commitment, ability to work in times of crisis and the ability to work out ways for management.

The water users groups still retain certain rules and regulations as set by the project but have largely been finding innovative ways for deep tubewell management. Rule making has been a larger challenge in case of the older tubewells when compared to the new ones. The design of the tubewells, in their designs, pump sizes and operating costs, dominates the way the farmers are able to devise and modify certain rules. The rules and regulations that they still maintain are those related to demand charge. This is because the payment of demand charge is cost incurred for a flat rate of electricity that is related to the capacity of the electric motor. They have made their own rules with respect to water allocation and distribution and for gathering monetary resources for the upkeep of the deep tubewells. Some of these innovative ideas are derived from the experience in kulo irrigation and some have been drawn up specifically to find out ways of bringing in cash for the deep tubewells. Each committee has created it own set of plans and rules for managing the tubewells. These include changing rules for water demand; setting fees for growing lentils or any crops on the canal bank; mobilising labour resources for deep tubewell cleaning; setting fees for transporting shallow groundwater through deep tubewell canals; having vouchers to reduce cheating; creating rules to control the installation of shallow tubewells, controlling the members of the committee by the executive itself; trying to build accountability. The cases show that farmers make use of different options to make and enforce the rules and regulations. They make use of their formal positions within the water users committee, the informal institution of the village, the formal wards and the kulo.

Notes

[1] Pradhan Pancha is the equivalent of the VDC Chairman in current times. In the Panchayat period (1961-1990), other political parties were banned.

[2] The farmers claim that the contractors lined the existing canals instead of constructing new ones.

[3] This is a special *puja* (prayers, offerings) done by the Tharus to ward off evil spirits away from the village. The *puja* is done in the *simana* (or the boundary) of the village in four directions. *Sibaan Puja* is done once every year. *Sibaan* in local terms means 'boundaries of the village'. The Nepali equivalent term is '*simana*'. This is performed before the paddy season. There is place called as the *dibaar* in every village. A plank of wood or bark of tree is dug into the ground. The villagers drive in iron nails in the ceremony to ward off evil spirits. Animals are also sacrificed in this ceremony. The money to perform the ceremony is collected by the *puja* committee. This fund is called as the '*sokha*' in Tharu terms.

[4] *Sadak* means road. And *sadkaai* means cleaning the village roads and cleaning and managing the drains.

[5] The conversion has been done from bighas to hectares. A bigha is the local unit of area measurement. 1.48 bighas equal one hectare.

[6] This was stated by the deep tubewell chairman. It could be verified by looking at the background of the members.

[7] All development activities are carried out through the 'participatory approach'. A multitude of groups are created in the local level in order to implement these activities. Besides these, other forms of group making exist in the form of dairy cooperative, savings and credit groups, temple reconstruction group etc.

[8] A new political party was formed in 2001 called the Rashtriya Samata Party from a leader from Myagdi district. Most of the people from this village belong to the same mother village and also come from the same ethnic group as the founder of the political party. It was formed by a breakaway faction from the Nepali Congress Party.

[9] Durga Puja is one of the biggest Hindu festivals dedicated to the goddess Durga.

[10] This is a very important temple in the area. It covers a large area and the village has made a committee for the temple. This committee is planning to link this temple with other villages and make a recreational park also. This site is of historical importance, as ancient archaeological brickworks have been discovered similar to the ones in Lumbini, the birth place of Buddha, also in Rupandehi.

[11] Exchange rate US$ to NRs. In 2004: IUS$=NRs. 73.67.

[12] In some tubewells, for example in the Kotihawa tubewell, the *Samiti* meets on the second of every month in the Nepali calendar.

[13] A meeting is sometimes even held in front of a village shop.

[14] One such area is called Bankatti and the other called Jabarjastapura. *Ban* in Nepali means forest and *kaatney* means to cut. So Bankatti means cutting the forest. The people cut down the forest, claimed land and settled down here. *Jabarjasti* in Nepali means 'forcefully'. *Pura* means settlement.

[15] Right-wing (monarchist) party.

[16] In July 2002, the government of Sher Bahadur Deuba dissolved all District Development Committees, municipalities and Village Development Committees which are the local elected bodies and replaced them with government employees. There has not been an election since that period. The posts of VDC chairmen and elected ward members mentioned in the chapter were effective only till that period.

[17] This system was initiated in Nepal by the *Chattis* and the *Sorha Mauja* Irrigation Systems. These farmer managed irrigation systems get a flood of visitors and consultants. Sometimes the visitors take long hours to discuss and visit the whole canal systems that take much time for the farmers, who have to take off time from their busy schedules.

5

Strategies for Better Water Control in Madhaulia

"At this moment, we are more concerned with having total control over the different water sources so that we will not be in the mess we are in today. After the handover of the deep tubewells, we have learnt a lesson. We should not trust anymore. So you see, until we have total control over Lausi we will not let go. Of course we will keep the rights to the Chattis and the Lausi, as well as to the deep tubewell."

- A farmer in Bihuli.

This chapter is a case study of Bihuli village in Madhaulia VDC, and its struggle for securing conjunctive use after the handover of the deep tubewell. In this chapter, I examine the different strategies and means employed by the farmers in the process of gaining control over both groundwater and surface water. This case study shows how the villages are capable of shifting between different institutions for water management and making their choices from various normative repertoires to suit their purposes at different points in time. This chapter shows that intervention processes are carried out in a dynamic context and that the inserted institutions are susceptible to changes and transformations due to different factors. It also shows that the intervention process itself can be an inducing factor in bringing about transformations. The management transfer process also initiated new forms of cost sharing mechanisms in groundwater irrigation between the farmers and sharecroppers and shaped the way choices were made between different sources of water.

History of water use in Bihuli

A deep tubewell was drilled in Bihuli village in 1983. The installation of the tubewell was completed in 1985. It was designed to irrigate a total of 145 hectares (BLGWIP, 1999) inside Bihuli and parts of Chamkipur village. The well is located in ward number nine of Madhaulia VDC but largely irrigates Bihuli village, which is in ward number eight.

Bihuli, had two different sources of surface water before the deep tubewell was installed. It was the last mauja of the Chattis Mauja irrigation system but also had access to Lausi Khola. The kulo from Lausi existed as an independent kulo. The farmers used to take water directly to the village by building a temporary intake in the stream. Lausi is a perennial stream which flows out of a subsurface spring from a VDC north of Bihuli. The stream flows through Madhaulia VDC and flows on to the urban areas of Bhairahawa in the south. After this, it flows across the border into India. Besides these two major sources of water, the village also had access to the jharan water that came up after the upstream villages irrigated.

After the initiation of the land registration process from 1981, the farmers had difficulty in making the brushwood dam in the Lausi Khola and had to abandon using water from this stream. The area near the intake was registered as private land. The landowners in the vicinity of the intake did not allow the farmers to cut the vegetation that was needed to make the dam. After the deep tubewell was installed, the farmers also stopped irrigating from the Chattis Mauja too. For several years they irrigated only from the deep tubewell.

The process of handover set off new sets of actions to increase water delivery. First and foremost, the farmers decided to redefine their water rights with Chattis Mauja. Second, they tried to find out different strategies to manage the deep tubewell, and evolved their own sets of rules and regulations. Meanwhile they reduced groundwater use. Third, they organised to start a process of application for a new dam on the Lausi, in order to have better control over surface sources. The following sections have been structured around the different processes that have led to the transformation of irrigation in Bihuli. The role of the different actors and their different strategies are analysed with respect to these processes.

FIGURE 5.1: Location of Bihuli and the Kulos to it from Lausi Khola and Chattis Mauja

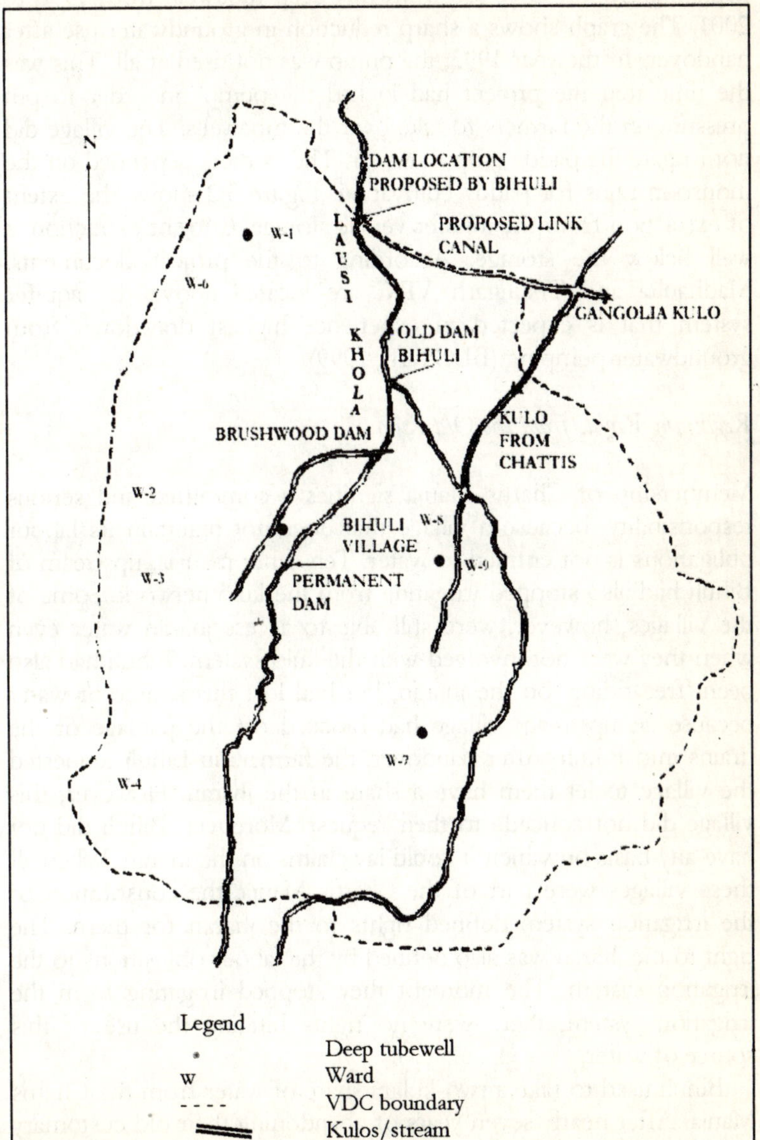

Source: Field work 2002/04

Figure 5.2 shows the different actions that the farmers undertook. The different actions and processes are set against the graph depicting groundwater use from this deep tubewell, from 1990 to 2001. The graph shows a sharp reduction in groundwater use after handover. In the year 1993, the pump was not used at all. This was the time that the project had locked the pumps in order to put pressure on the farmers to take over the tubewells. The village did not irrigate the paddy crop that year. The farmers depended on the monsoon rains for paddy cultivation. Figure 5.2 shows the extent of extraction from the aquifer versus storage. Current extraction is well below the storage. According to the project documents, Madhaulia and Tikuligarh VDC are located above the aquifer system that is expected to experience highest drawdown from groundwater pumping (BLGWIP, 1999).

Regaining Rights from the Old Kulo

Membership of Chattis Mauja signifies a committed and serious responsibility, because a village that does not maintain its labour obligations is not entitled to water. Ten other maujas, upstream of Bihuli had also stopped irrigating from the kulo network. Some of the villages, however, were still able to access jharan water even when they were not involved with the kulo system. Bihuli had also been 'free riding' on the jharan, but had lost this source of water because an upstream village had blocked off the passage of the drains into Bihuli. After handover, the farmers in Bihuli requested the village to let them have a share in the jharan. However, this village did not concede to their request. Moreover, Bihuli did not have any basis on which it could lay claims on the jharan. When all these villages were part of the Chattis Mauja, the constitution of the irrigation system defined rights to the jharan for them. The right to the jharan was also defined by the labour obligations to the irrigation system. The moment they stopped irrigating from the irrigation system, there were no rights binding the use of this source of water.

Bihuli used to take a two-kulara share of water from the Chattis Mauja. After nearly seven years of abandoning their old customary rights to the irrigation systems and converting to groundwater irrigation, the farmers decided to renew their rights with it again. The new constitution of the Chattis Mauja had officially removed

FIGURE 5.2: Actions taken by Bihulians to gain control over different sources of water

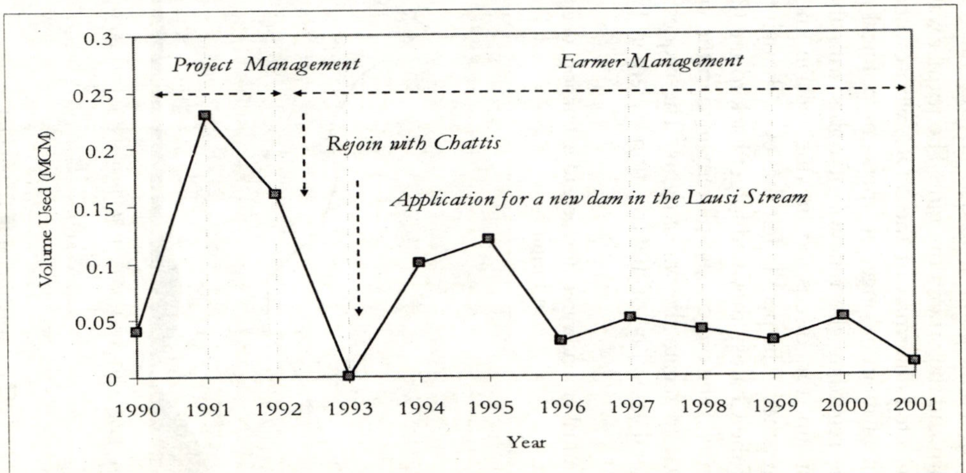

Source:

a) *Field work 2002/04 for the actions and the processes*

b) *The deep tubewell water use records (1990-2001) and BLGWIP, 2002.*

these villages from the list of maujas. In order to renew the rights, the mauja had to put a request to the main committee of the irrigation system.

A village meeting was set up under the leadership of two persons. These were the political leaders in the village. One of them was a Tharu and the other a migrant. The former was also the democratically elected chairman of the VDC, who owned land in this village. Both of them belonged to the same political party, the UML. The issue regarding the renewal of rights with the Chattis Mauja was put up for discussion in the village meeting. A consensus was reached. The people felt that even though they were in the tailend of the Chattis, 'they had to get back to the Chattis for the sake of security'. The old canals still remained in the village and the only step was to regain the water rights. The request was put forth to the general assembly[1] of the irrigation system in 1992.

FIGURE 5.3: Annual groundwater storage and extraction, deep tubewell in Bihuli

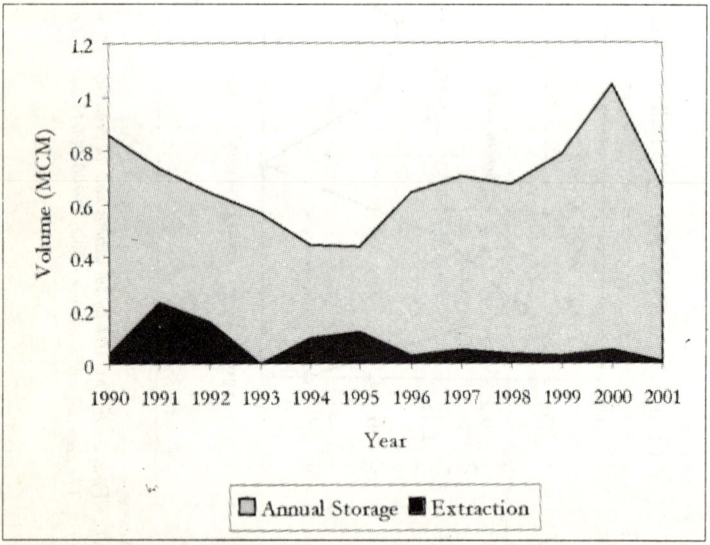

Source: BLGWIP, 2002

A provision exists in the written constitution of Chattis Mauja, that allows villages to open new outlets, leave their rights to the irrigation system. There are also provisions to change the location of outlets. According to the constitution, any village that wishes to open a new outlet has to put forth the request for discussion in the general assembly (Chattis Mauja constitution versions 1983, 1993, 1995). This request has to be endorsed by the majority in this assembly. There is also a provision for rejoining. The farmers in Bihuli took this rule as a reference and made their justifications to claim their water rights back to the irrigation system.

Each village that wishes to rejoin has to pay a fee to the irrigation system. The rate fixed by the constitution of 1989 was an amount of NRs. 10000[2] for one kulara of water. This right also entailed rights to the jharan water. In addition to rules for joining the system, there are provisions for leaving the irrigation system. Any Mauja wishing to leave the irrigation system, is obliged to pay a sum of NRs. 12000 to the main Chattis Mauja[3].

The owners of two hundred hectares of land inside Bihuli and Chamkipur applied for renewal of their old water rights. The money was collected on the basis of bighatti (in proportion to area) from 193 households in Bihuli and thirteen households in Chamkipur. Previously, the village used to take two kulara share. This time, they reduced it to one kulara. One of the reasons for reducing the share was their inability to guarantee labour when needed. The opening of thirteen small scale industries[4] in this VDC had directed labour in agriculture to the higher paying factory jobs.

After the mass meeting[5] endorsed the request of Bihuli Village, they were officially allowed to become a part of the irrigation system again. The upstream villages had not yet joined back with the Chattis. After Bihuli officially joined the irrigation, the village immediately upstream of it had to succumb by opening the passage of the jharan to Bihuli. The following year, this village too renewed its surface water rights. One by one the rest of the villages that had abandoned their surface water rights rejoined with the Chattis Mauja again.

By the time the tubewells were handed over to the farmers, the power relations in Bihuli had definitely shifted towards the hill migrants. Even though Tharus make up the majority, the hill migrants are more visible in decision-making processes in the village. The shifts from the deep tubewell to conjunctive use were made been possible because of the roles of some actors in this village.

Several actors figure prominently in decision making with respect to water resources in Bihuli, and one of them is chairman of the deep tubewell. He has been chairman of the tubewell right from the time the deep tubewell was handed over. He is also one of the main figures involved in decision-making processes in the village as well as the main person involved in the process of negotiating with the old kulo. S.Sharma, took early retirement from his job as government officer and came to live in the village. He owns around fourteen hectares of land between his family and also oversees the management of the land for the previous landlord in the village, who does not live in the village. Sharma started as a government officer but is now a full-fledged politician and a prospective member of parliament from the UML party. One of the first acts of responsibility that he took over in the village decision-making process was as chairman of the deep tubewell. He took the leadership to negotiate with the irrigation system in order to renew the water rights. He was not alone in this endeavour. The chairman of the VDC who was also from the same village, was also involved. His role was more in the capacity of an elected leader. The villagers regard Sharma as the main negotiator. Rules for renewal did exist in the constitution. But it had to be worked out, interpreted and negotiated. He used his contacts with the executive committee of the irrigation system in order to work it out. He has a very strong network based on political connections. The members of the water users committee agree that 'we try not to work on party lines, but finally it comes up to that (political networking)'.

After being successful in the process of renewing the rights to the Chattis, the chairman of the deep tubewell also became chairman (mukhtiyar) of the revived kulo in Bihuli village. He has developed strong contacts with the Chattis Mauja network in his role as the chairman of the village kulo and in his role as a member of the political party. After serving two terms as the kulo chairman, he decided to quit the job. Responsibility was then shifted from Sharma to a new migrant with a long experience in the army. Later on, Sharma rose to the ranks of treasurer of the Chattis Mauja main committee. The executive committee of the irrigation system consists of a chairman, a vice-chairman, joint-secretary and nine regional members. These twelve members nominate three other members[6] from within the command area. One person is chosen as a treasurer.

This village, which had left the membership of the Chattis for many years, is now linked up with the mainstream irrigation development of the district through different networks that such actors have carved out for themselves. Besides being the treasurer of the Chattis, Sharma is also the treasurer of joint committee of Sorha and Chattis Mauja. This is the main committee that deals with the water distribution from the Tinau River between the two kulos: Chattis Mauja and the Sorha Mauja. He was also nominated to be the chairman of the Rupandehi District National Irrigation Committee or the Rupandehi component of the National Federation of Irrigation Water Users' Nepal NFIWUAN[7]. In his role as the chairman of the Rupandehi component of the NFIWUAN, he is also involved in bringing the issues related to groundwater to policy makers. There is a network between him and the leaders in Tikuligarh and another VDC that is working towards building a stronger coalition of the deep tubewells. As he is the chairman of the district committee and represents the different water users' association inside the district. Fifty-three water users associations from different irrigation systems in the district are members of the district level irrigation committee (NFIWUAN, 2004).

One of the recent endeavours in several deep tubewell areas has been efforts at reorganisation, so as to form a pressure group. Even though the BLGWIP is classified in irrigation policy circles as a large-scale project, all deep tubewells converted into small irrigation systems after handover. One of the major focuses of the group is to have the demand charge for tubewells eliminated. Besides the 'irrigation groupings' this is also a strong group of individuals sharing a common political ideology. These actors did not make use of the deep tubewell co-ordinating committee, set up by the BLGWIP before handover, but instead started a new process of setting up another committee. The chairman of Tikuligarh VDC was chosen as a representative to the district level committee. This was one strategy to represent the deep tubewells in the national federation.

Going back to the issue of Chattis: what is interesting here is that, Sharma was successful in the negotiations and he did get the recognition and support for leading the negotiations. In the eyes of the other villagers, he has made a place for himself as some sort of 'provider'. However, the very fact that the FMIS had the rules for adjustments is very important. It was easier to get back the rights because defined rights of renewal existed within the FMIS constitution. If such a rule had not existed, the process would have been longer and more difficult. This shows the flexible nature of the irrigation system in incorporating different types of uncertainties and changes. Eleven maujas were given back their rights even after leaving the irrigation system for many years. This is a substantial political arena for district level politics.

Evolution of the deep tubewell committee

Two different processes were undertaken to support the deep tubewells after handover. First of all, the water users committee worked out different strategies to manage the deep tubewell on their own. In recent years, they are involved in networking with other deep tubewell committees in other VDCs to garner support for deep tubewell irrigation, as mentioned above.

The first problem that the management committee faced was the same one as all other deep tubewells faced. This involved the issue of collection of demand charge. The farmers in the low lying areas inside the command area refused to pay the demand charge, because they did not use the tubewells. In the first year after taking over, the committee made it compulsory for all farmers to pay the fee, irrespective of the fact whether they used groundwater or not. They set the rate at NRs. 735 per hectare. This strategy was worked out in order to set up the initial fund to manage the deep tubewell. In the subsequent year, they reduced the rate even when the area under irrigation decreased when the farmers in the low lands refused to pay. The demand charge was set at NRs. 662 per hectare in 2004 and the operating cost was NRs. 180 per hour. The committee collected the demand charge only for 34 hectares. This means, only 24 percent of the designed command area is functional. The salary for the pump operator was set at only NRs. 5,000 per year. This is a very nominal sum, and in order to supplement it, the committee decided to let the pump operator irrigate his plot without any charges.

The committee then set rules for distribution of the free flowing artesian water from the deep tubewell. The discharge obtained from the flowing artesian source is slightly more than the average discharge from a shallow tubewell in this village. This water was not accounted for before handover. In 2002, the rate for the water was set up at NRs. 20 per hour. In 2004, it was raised to NRs. 30 per hour. This flow is available from the monsoon up to winter season. A rotation is set up for distributing this water for the winter crop.

The cleaning of the deep tubewell canals has been set up as

regular activity before major irrigation times. This is carried out in the dry season preceding paddy cultivation and the wheat season. The committee has also made a rule that allows them to sell the wood from the trees around the pump house in order to increase their funds. There are four fishponds in the village. These are mostly in the water- logged areas but need to be replenished in the dry season. A rate has been set up for filling these ponds. A farmer who wishes to use water for the ponds has to pay a rate of NRs. 240 per hour.

After taking over the deep tubewell, the committee fixed the minimum time that a person could request for water. It was set at six minutes. This was not easy to put in practice as it was not practical to open the motor for such a short time. The pump is opened for six minutes only when the committee feels that irrigation is critical. The committee encourages farmers to come in groups of at least two or three to request for water.

The pump has broken down once since it was handed over. The BLGWIP bore the cost for it. The project bore such costs for two years after transfer. The coordination for other types of assistance from the project is done by the chairman in coordination with the regional coordinating committees as mentioned in chapter four. The committee has not made any rules for organising funds for large scale repair and maintenance operations. This is true for all the deep tubewells. Farmer strategies are more concentrated in finding out the means by which they can find monetary resources for the tubewells from other sources. It is also not clear how costs will be shared between the farmers and also between landowners and the sharecroppers.

A new set of dynamics was emerging in the year 2004. The local-level government bodies had been dissolved since August 2002. The elected members of the ward and the VDCs were not in a position to work or make official decisions at the local level. The VDC was administered by the secretary. The secretary is an administrative position in the VDC while the rest of the members consist of locally elected politicians. The secretary was observed to be in constant contact with the local political leaders because he had to take suggestions from them to implement certain activities. Likewise, due to the absence of the ward as a government unit, the responsibility for certain activities that was normally not allotted to the kulo committee was being handed over to it. The kulo committee has always been responsible for irrigation activities and other activities specifically related to the pujas done by the Tharus, as well as the traditional cleaning and maintenance of the village roads. The ward as the unit of government has the right to spend the development budget allotted by the government. However, after the wards were dissolved, some of the functions of the ward were transferred to the kulo committee. The kulo committee at the village level consists of a chairman, a secretary and a chowkidar. The secretary of the kulo is responsible for maintaining the accounts related to irrigation. In the absence of elected members in the ward, the people of Bihuli agreed to grant the kulo committee the rights to oversee and use the budget allotted to the ward by the government. The committee organised activities related to road maintenance. The secretary of the kulo also maintained a ledger book where he accounted the income and expenditure inside the ward. He collected fines both for irrigation as well as for other social activities. The account is then audited by individuals chosen by the village and is verified and signed by them. The fine for irrigation, also called the khara, is the fine that the farmers have to pay for not attending canal cleaning operations inside the village. There are different types of fines collected in the village. One of these is a fine for social indiscipline. Even though such fines were previously not being spent for irrigation, one incident in 2004 shows how farmers have been trying to draw funds for groundwater from different domains of social life.

There was to be a wedding ceremony in the village of a Tharu girl with a Tharu boy from another village. The bride eloped with another (Tharu) boy from the same village, just before the wedding day. The family was embarrassed and a meeting was called in the village. After much discussion they decided that they had to somehow punish the village boy for creating embarrassment to the family of the girl in front of the would-be bridegroom. The family of the would-be groom was asked the extent of loss they had incurred for the arranging the wedding. The family had already paid NRs. 5000 for the wedding band and some other items. The village leaders decided to fine the boy who had run away with the girl from the village. The boy paid a sum of NRs. 3500 to the family. The remaining amount of NRs. 1500 was put in the village fund for activities including groundwater.

Cost sharing in irrigation

Altogether six families give out land for sharecropping in the village. Around twenty percent of the total area of the village is under sharecropping. The land distribution pattern in this village is skewed with few large landowners holding most of this twenty percent of the land. The common sharecropping contract is the so called fifty-fifty share for both inputs and harvest. Cost-sharing involves sharing the costs of inputs and harvest for both monsoon crop as well as winter crops: wheat and mustard. The straw is also shared equally between the landlord and the sharecropper.

Cost sharing of inputs involves costs incurred in fertilizer and insecticides. Seeds are kept as a share from the produce. It is supposed to be shared equally. This depends on the relation between the owner and the cultivator. All costs for land preparation including human labour and tractor costs are borne by the cultivator. The cost for irrigation has also evolved with respect to the changes in choice of source of water. In case of the kulo, the kulahi is done by the cultivator. No accounting is done for this labour, as it comes within similar costs for labour in other activities for production. For farmers using deep tubewells, the landowner pays the demand charge. This system of sharing costs for irrigation came up in Bihuli after handover. The running charge or the cost involved in actual irrigation is borne totally by the cultivator. It is not yet clear who will pay for costs incurred in case of major breakdown (the people feel it should be the landowner). The landowner is required to pay cash required for canal maintenance. However, this is true only for those farmers who use both surface water and groundwater. Some farmers, who use only groundwater, divide the cost for demand charge. Table 5.1 and 5.2 show the cost involved in paddy and wheat production for one hectare of land, and the system of cost-sharing between the landlord and the sharecropper. The tables show that the main costs are borne by the cultivator. Cost for surface irrigation is borne entirely by the cultivator. Management transfer of deep tubewells has shifted one more burden of input cost to the cultivator. Even though large landowners were very active in the process of resisting handover, their share of cost in irrigation is still relatively low when they give out land in sharecropping. In villages where land distribution is skewed, it is not easy to make regulations as regards joint use of groundwater and surface water.

TABLE 5.1: Cost of input and cost sharing for paddy cultivation (2001)

Paddy	Rate			Cost Sharing (NRs.)	
	Nos.	NRs/day	NRs.	Owner	Cultivator
Labor (human)					
Paddy nursery bed	5	100	500		
Broadcast	5	60	300		
Seedlings separate	40	60	2400		
Transplanting	30	100	3000		
Weeding	25	60	1500		
Cutting	18	60	1080		
Lifting	10	100	1000		
Threshing	10	100	1000		
Tying straw bundles	5	60	300		
	148		*11080*		11080
Tractor					
Land preparation		NRs/hr			
Bed	0.33	360	119		
First plough	3	360	1080		
Second plough	3	360	1080		
Third plough	3	360	1080		

Threshing	3	360	1080		
	12.33		*4439*		4439
Fertilizer		NRs/kg			
DAP	40	22	880	440	440
Urea	40	12.5	500	250	250
Muriate of Potash	10	15	150	75	75
Irrigation					
Demand Charge DTW (NRs. 450 per bigha)			450	450	
Running Charge DTW (NRs . 180 per hour)			1620		1620
			3600	1215	2385
Total cost per *bigha*			19119		
Cost sharing (NRs/*bigha*)				1215	17904
Cost sharing (NRs./hectare)				1787	26329
Yield (quintals)					
	25				
Farm gate price (NRs/quintal)	650				
Total selling price (NRs.)	16250				
Net benefit per bigha	-2869			15035	-1654
Net benefit per hectare	-4219			22110	-2432

Source: Field Work 2001/02

TABLE 5.2: Cost of input and cost sharing for wheat cultivation.

Wheat				Cost Sharing (NRs.)	
	Numbers	Rate (NRs/day)		Owner	Cultivator
Labor (human)					
Land preparation	2	60	120		
Seeding	1	60	60		
Harvesting/cutting	10	60	600		
Carrying	10	60	600		
Threshing	4	60	240		
			1620		1620
Machine (hrs. of use)					
	NRs/hr				
Tractor (land preparation)	4	360	1440		
Tractor (threshing)	3	360	1080		
			2520		2520
Other Input (kg)					
	NRs/kg				
Seed	80	22.5	1800	900	900

Fertilizer					
DAP	100	22	2200	1100	1100
Urea	100	12.5	1,250	625	625
Muriate of Potash	50	15	750	375	375
Irrigation			1440		1440
				3000	4440
Total cost per *bigha* (NRs.)			*7440*		
Cost sharing per *bigha* (NRs.)				3000	8580
Cost sharing per hectare (NRs)				5074	13059
Yield (q)	10				
Farm gate price (NRs/quintal)	700				
Sale of husks (NRs/q)	700				
Total selling price (NRs.)	7700				
Net benefit (NRs.)/bigha	260			4250	-1180
Net benefit (NRs.)/ha	382			6250	-1735

Source: Field Work 2001/02

Struggles to Get More Surface Water

The previous sections dealt with how conjunctive use complex emerged in Bihuli. The following sections deal with how the farmers in the village are strategizing to increase the surface water supply into the village. The focus is now on the struggles around Lausi. The interactions involve several villages inside and outside the VDC, that are hydrologically connected to the Lausi stream. This source of water, which had been ignored in the process of implementation of the deep tubewells, has brought about tremendous changesin irrigation management practices in Madhaulia VDC. The BLGWIP was implemented within pre-defined boundaries and did not take into consideration the diversity and the presence of these 'small' sources of water. The farmers later reduced groundwater use dramatically and Lausi became the focus of contestation inside Madhaulia. Regaining the water rights from the Chattis did not give the people in Bihuli village a full sense of security. They were still in the tailend of the Chattis Mauja. Lausi, on the other hand ensured them stronger water rights.

Several villages in Madhaulia VDC have been irrigating from the Lausi Khola. There are three different intakes inside Madhaulia VDC. The first intake is that of Bihuli village. The second is a brushwood dam, from where the farmers in ward number two and three had historically been taking water from the Lausi. The third is a permanent dam constructed by the District Irrigation Office (DIO) (see Box 5.1). This dam irrigates the lower portions of the VDC. Another village also irrigates from the stream. It is the village next to the northern border of Bihuli. This village is Gangobalia, in Gangobalia VDC. The different kulos used to divide the water in the stream by means of a wooden proportioning device called a *saacho* and the share of each village was 75 haath[8] of water.

The farmers in Bihuli made several attempts to get back water from Lausi. They filed a formal request for a dam with the district irrigation office in 1993 under its farmer managed irrigation systems rehabilitation programme. The department of irrigation undertook such programmes as the Irrigation Line of Credit (ILC) and the Nepal Irrigation Sector Project (ISP)[9]. The farmers in other wards had also obtained the permanent dam through the same implementation programme (see Box 5.1).

BOX 5.1 From conjunctive use to kulos

The villages in ward number five, seven and nine in Madhaulia VDC, along with ward number two of Gangobalia VDC, applied jointly for a permanent dam in the Lausi. The application for the dam was made in 1997 at the DIO. This was made possible through the Irrigation Line of Credit Program of the WB/DOI that supports the rehabilitation of FMISs. The design command area under this is 320 hectares. This area also falls under parts of command area of several deep tubewells.

The project was approved in 1998 and the construction started in 1999. It was completed in 2000. The application process for the participatory approach requires a registered WUA. The farmers did not have a written constitution or a registered WUA. A formal water users group was then registered. The participatory approach required the farmers to pay 25 percent of the total cost of the irrigation project. They were required to contribute only 2.5% as cash deposit, which was NRs. 42,000 while the rest was met by contributing labour. Money was collected from the farmers at the rate of NRs. 18 per hectare. Even though this is a very low amount, some farmers cultivating a total of 34 hectares of land were reluctant to pay. These farmers still use the deep tubewell. They requested the WUA to be a part of the irrigation system in 2003. In 2004, the WUA was considering their request. They would however, have to pay the amount that the rest of the farmers had paid for during the initial construction.

The farmers in ward number nine have stopped paying the demand charge for the deep tubewell. Only the chairman of the kulo was still paying the demand charge for the tubewell. The deep tubewell is located in ward number eight. The water users group has made a separate rule for the people in ward number nine. Farmers in ward number nine could still get deep tubewell water provided they agreed to pay NRs. 100 more than the rest of the farmers in ward number eight who have been paying the demand charge regularly. Ward number nine was still asssured of groundwater lest they need it for the future. The chairman of the kulo does not pay the total demand charge. He only pays for a small part of this land. He sometimes prefers to use groundwater in the dry season, as there is a rotation for water from the kulo.

Source: Field work: 2001/04

Bihuli made a formal request to the district irrigation office in 1993. However, this request has not yet materialised even after so many years. One of the main reasons is because this request led to inter-village conflict inside Madhaulia VDC and the farmers in other wards filed a complaint against the application at the VDC. The farmers in Bihuli on the other hand did not give up and followed this process while trying other measures at the same time. They have been strategising for both temporary as well as long term measures.

The Bihulians invested in stones and gabion boxes and had a wooden aqueduct built at the point where the kulo of the Lausi crossed the kulo from Chattis. In order to gather the funds for this, they made use of the PDDP program that was being implemented in the villages till 2003. The programme was called Implementing Priority Productive Investments phase in which, the COs or Community Organisation finalizes the Village Development Plan (VDC Plan) for productive infrastructure priorities. Support is offered through the Seed Grant Fund (SGF) to fund COs' local investment in small-scale productive infrastructure development such as irrigation projects, drinking water supply schemes, micro-hydro, development of higher levels of cooperative enterprises, commercial forestry, and others. There are four savings groups in Bihuli formed by the PDDP: The Janabikash, Navadurga, Lali Gurans and Sahasi Mahila. These are community organizations (Cos) formed by the PDDP programme in the VDC. The total estimate for the material needed at the headworks was estimated at NRs. 108970 by the programme technician. The VDC put in NRs. 17500 and the district development committee (DDC) put in a matching fund of NRs. 17500. The rest of the money had to be collected from the farmers and was NRs. 27243. The village made use of the groups to access the programme funds, but did not collect the remaining money entirely from the savings groups. They pooled in a certain amount of money from the village income (from different types of fines) into the fund. The rest of the money was collected on the basis of bighatti[10]. After successful collection of the required amount of money, they organized again to form an alliance between four COs to approach the programme. In order to access the PDDP fund, it was necessary to approach through the COs. All COs are different and have their own ideals and are not necessary politically neutral. Even though neutrality is the so called objective while forming COs, each group is formed to achieve certain objectives. The COs crossed individual group boundaries[11] in order to work out a strategy that benefited the larger village society. Despite this attempt, the village was still not able to construct a proper intake.

Forming and dropping alliances

As has been mentioned above, the application made by the farmers in Bihuli was pending till the summer of 2004, because it had been strongly opposed by the farmers in adjoining villages. The people of ward number two and three had filed a complaint in the VDC expressing their opposition to this strategy. These are the original old settlements inside Madhaulia VDC. Historically, these villages were irrigated through a network of four kulos: Paangoda, Semarhwa, Gointadi and Jharnerwa. This small irrigation system irrigates a total of 120 hectares, inside the VDC. The farmers were opposed to the Bihuli dam, because they feared that, if it is built in the location that Bihuli had proposed their area would be affected. They claimed that one of the kulos in their network would not get enough water and another part would be submerged. The farmers in this kulo system are very proud of the kulo and strongly defend it as being even older then Chattis Mauja itself because some parts of the village are old Tharu settlements. This kulo system has its own organisation[12]. Groundwater was supplied to these wards from the deep tubewell installed in ward number six. However, they did not use it anymore and were very adamant on protecting the surface water flow into their village.

The farmers in Bihuli, on the other hand, who assert equal rights on the kulo. had formed an alliance with another village called Babani in Tikuligarh VDC in order to form their own kulo network. They later abandoned this strategy, and formed another alliance with other villages in other VDCs. The new water users group consisted of Bihuli, a village in Gangobalia and another village in adjoining Karaiya. There is a very important reason for Bihuli to change alliances. They wanted to shift the position of the intake to another location, higher up the stream. In order to do accomplish this, they mobilized the farmers in Gangobalia village in Gangobalia VDC. The people in Gangobalia village agreed to be part of this new alliance, because they too wanted to improve water supply into their village. Gangobalia, like Bihuli was struggling to retain its deep tubewell.

The farmers in Bihuli had to work out a new way of bringing water into the village through this new intake. Changing the intake to a location higher up meant, that they had to construct a new canal. They do not plan to construct a brand new canal. The existing Gangobalia kulo runs at the border of the two villages. The farmers in Bihuli planned to bring the water first through the kulo from Gangobalia and then transport it through a short new link-canal to the old existing kulo. The water would be divided between the different villages from that point onwards. The water users' association submitted NRs. 10000 as part of the 'farmer participation component' for the new irrigation system. Rehabilitation works in farmer managed irrigation system are implemented through a participatory approach that requires the water users' to contribute 25 percent of the total cost estimated. The payment was made in 2001 (Source: ledger book of the Bihuli irrigation committee). The decision to actually implement the project from the side of the government had still not been made by the start of the summer of 2004. The existing chairman of the Bihuli kulo has been given the responsibility of managing the future Lausi kulo.

It has taken the village many years for their project to materialize. One of the main opponents to this project was the former chairman of the VDC himself, a resident of the village that opposed the dam. He justified his action because it was not fair for him to make a decision that would benefit only a part of his jurisdiction, while the other part was adversely affected. In his role as VDC chairman he did not want to make controversial decisions inside his own political arena. The main actor behind the dam initiative was a small politician in Bihuli who belonged to the same political party as the chairman. The more powerful leaders in Bihuli were not directly involved in this endeavour. There was much dissatisfaction between the small politician in Bihuli and the 'bigger' leaders belonging to the same political party inside Madhaulia. The farmers in Bihuli argued that their case had become stronger after they made the new strategy to involve Gangobalia. Gangobalia does not belong to Madhaulia VDC. Therefore the power to make decisions was shifted between two VDCs. In addition, it was also the older user of the Lausi and had strong water rights. This would make it difficult for ward numbers two and three in Madhaulia to object to the process. The farmers in Bihuli therefore, made use of different normative repertoires to base their claims on Lausi. The government intervention programme offered them an opportunity (as it did for some village downstream) to strengthen their water rights. They shop around for the best possible way to strengthen their claim over the water by choosing and dropping different alliances. For instance, the choice of Gangobalia provided them a basis of stronger water rights but also another forum: the VDC to strengthen their case.

Conclusions: Shifting between institutions

The farmers in Bihuli have been choosing, rejecting, manipulating and manoeuvring between kulos and the deep tubewell. This case illustrates that institutions are dynamic and susceptible to changes and transformation, and that changes in the use of one source of water affects the management practices around the other. The course irrigation will ultimately take depends on the agency of the farmers. They are actively engaged in the process of adjusting and incorporating to different types of uncertainties and events. Groundwater provided the farmers with security so that they abandoned their surface water rights. The transfer of the tubewells, set off different processes of adjustments for irrigation. It involved a revived interest in kulo irrigation. Bihuli, along with ten other maujas had left the surface irrigation but later joined it again after several years. This sort of movement between common property resources like the kulo and deep groundwater affects the technological and organizational performance of both the irrigation systems.

Management transfer of the deep tubewells did not involve only the transfer of 'irrigation management tasks' in groundwater irrigation. It set forth a dynamic whereby the farmers started getting actively engaged in strategic planning that would help provide more security to them in irrigation. They are thus involved in larger governance issues around water in their village. 'Collective security' (Vincent, 1995), became the objective of the irrigators after management transfer in order to complement the technological constraint provided by deep tubewell irrigation. The leaders who make decisions shy away from making any definite rules for operation and maintenance of the deep tubewells. It is clear from the case study that in order to meet the cash requirements, they work out strategies in such a way that the farmers do not have to pay directly. The example of pooling in village fines for irrigation (both groundwater and surface water) is a unique case which distinctly shows the blurring of the boundaries between water users group and the village as an institution and the evolution of new forms of governance emerging between the new groundwater technology and the existing social and political institutions.

An analysis of the cost sharing mechanism in sharecropping indicates how management transfer of deep tubewells shifted one more burden of input cost to the sharecropper. The cost of operation is divided to the different sharecroppers who still continue with groundwater irrigation.

Farmers are very actively engaged in using different options, strategic networks and different alliances in order to reach their goal. They make use of different legal systems that exist within their arena: the rules of the kulo, the options provided by the government intervention programmes in surface irrigation and also groundwater irrigation. They shop around, make alliances and also drop alliances in order to work out the best possible ways for them to strengthen their control over the various sources of water. The different actors are involved in multiple networks that have a common objective of increasing water security in the village. Even though they seem disconnected, they are connected through a common purpose and through the same actors. These networks are also used to develop and establish political and economic support from outside and as a means to establish a political base for the actors involved in the process. The networks that they develop for negotiating support from the government authorities link this village and VDC to the actors in Tikuligarh and other VDCs and at the district level. Khanal (2003), in his study on the processes and outcomes of irrigation management transfer in three surface irrigation projects in the Terai predicts that the individual water users' associations (WUA) 'will continue to develop economic and political network amongst themselves as well as other political institutions'. He gathers that they do so in order to find external support as well as to gain credibility at the local level. The findings in this study are in conformance with the findings of Khanal. His study was based on the transfer process of surface irrigation systems. A similar process of networking is emerging in deep tubewell irrigation. The important role that local leaders play in making choices in the technology and institution comes out very prominently.

According to Pradhan and Adhikary (2000), irrigation studies on water rights have focused more on use rights than in decision making rights even though property rights literature has defined different kinds of rights in water (rights to use, regulate, control and make decisions (Benda-Beckmann et al, 1996,1997 in Pradhan, 2000). In this case study, the decision to choose a certain source of water or to be a part of a certain institution is very important and entails acceptance of a specific bundle of rights and responsibilities. Moreover, the people who take up the leadership to make such decisions and the factors that shape their decisions are very crucial because they influence the future of irrigation water management between different sources of water. The local leaders make decisions on how they manage funds in the village and how a village is 'governed' in times of political uncertainty and no formal government structure. They are also the authority that have been what Khanal (2003) calls 'delegated authority' for local governance through the management transfer process. Therefore an analysis of their roles in how they help to bring about sociotechnical change in irrigation is very important.

Notes

[1] General assembly: This is made up of the main committee as well as the kulo chairman of each mauja or their representatives.

[2] The rate was NRs. 10000 per kulara in 1989. The 1995 constitution rated it as NRs. 20000. The rate in 2004 was NRs. 12000: it had been reduced but had not been published in the revised version of the constitution. This information was obtained from the treasurer of the kulo system in 2004.

[3] Any Mauja wishing to leave the kulahi has to pay a sum of NRs. 12000. However, it is found that no village really decides to leave their rights formally. They slowly start ignoring going to the kulahi.

[4] Factories include works for thread, iron rods, welding rods, chocolate, and pharmaceuticals.

[5] The mass meeting of Chattis Mauja refers to the meeting of the members of the irrigation system. Each kulara is entitled to four representatives.

[6] Section 2, sub-section 7 (ga). Chattis Kula pani Samiti ko Bidhan 2051.

[7] NFIWUAN

[8] A haath literally means one hand. This was the local measurement of water indicating the length from the elbows to the tip of the second

finger. Source: kulo chairman Lausi kulo (2).

[9] The Nepal Irrigation Sector Project (NISP) is implemented in forty districts of western and mid-western as well as far western regions of the country. This is done under the loan of the World Bank, a matching fund on the part of the government and farmer's participation. It started as a follow up program to the Irrigation Line of Credit Pilot Project, with loan assistance from the IDA. Rehabilitation and improvement of small and medium scale farmer-owned and managed irrigation schemes (both surface and groundwater) system improvement for farmer turnover of large irrigation schemes, infrastructure support in terms of farm roads, and river bank protection are the main components.

[10] The bigha is equivalent to 0.68 hectares. Bighatti refers to the collection of money on per bigha basis.

[11] Other saving groups existing in the village are those of the Nirdhan Banking (banking for the very poor), *Sana Kisaan* (Small Farmers Development Program and the PDDP. For larger loans, the farmers go to the Agricultural Development Bank. The PDDP has now phased out. Under the Local Governance Program or the LGP as it is called, the programme is now called as the Village Development Programme rather than PDDP. However it is understood that because of the multiplicity of groups existing in the villages, all development programmes are now to enter the village through the existing groups that have been formed by the PDDP. There is however, no longer the Seed Grant Fund anymore. Recently an NGO has invested in a maternity house in one of the wards in Madhaulia VDC. The cost sharing was done on the basis of 10 percent by the VDC, 10 percent by the DDC and 5 percent by the COs which covers the maintenance cost. The NGO bore the rest 80 percent of the total cost.

[12]It takes them three days to build the headworks. Each household is expected to participate on all three days. The canal cleaning works is done on the basis of land holding. All farmers who have less than 2.7 hectares are required to contribute three days in a year for cleaning the canals.

6

Shallow Groundwater Use in Mahuwari

"Oh yes, there was a kulo here. Look at this path that we are walking on. Do you see the path that the tractor is coming from? Well that is the kulo. But we do not use it anymore" Gaya remarked as he started his pumpset to draw water from the river right into the remnants of a part of the old kulo to irrigate the banana plantation. "Why did we stop using it? The river cut off a portion of our village and a portion of the kulo. We can use it if we want to because we can still connect it to the other village upstream, but you see, the whole village is already using tubewells. Nobody is interested in using it again…"

The farmers in Mahuwari village started irrigating from shallow tubewells since the last two decades. Mahuwari lies in Hatti Bangain VDC, and was never a part of BLGWIP. It was irrigated by a kulo, but later became a shallow groundwater irrigated area. This VDC, unlike other sites, has had less or no role in groundwater irrigation. The only exception was when a recent community development programme brought in a group shallow tubewell in the village. In this chapter, I examine the transformation of irrigation in Mahuwari village: the factors shaping the choices of the farmers for shallow groundwater over surface sources, how they try to gain control over it, the strategies developed in the process and the types of relationships that emerge around it. The chapter shows that the possibilities of profitable agriculture have shaped the choices for shallow groundwater over surface sources.

The first section gives a short comparison of water use in Mahuwari where farmers had access to only shallow tubewells and kulos and the other study sites where the farmers had the options of using deep tubewells also. In the second section, the background and socioeconomic status of the farmers in Mahuwari are described because this defines the way they make their choices in agriculture and therefore in irrigation. In the third section, I discuss the processes that have led the farmers in this village to make changes in their cropping pattern, and how this relates to the increased contestation in shallow groundwater and the movement away from surface sources. The different strategies the farmers in Mahuwari have been devising to gain more control over shallow groundwater individually and as a group is discussed in the next part.

Shallow Groundwater Use in the Research Area

The village of Mahuwari lies downstream of Tikuligarh VDC, separated from it by Chihiliya VDC. It lies between Tinau River and Ghagra Nala. Unlike farmers in Tikuligarh VDC, farmers in Mahuwari are not involved in surface irrigation. Most farmers who were involved in agriculture in 2004 in Mahuwari, did not have any experience with the kulo that existed there. (See Box 6.1). There have also not been any attempts by the village to revive the kulo and the farmers are more interested in shallow groundwater irrigation.

Shallow tubewell use was discussed in chapter four for Tikuligarh VDC, where farmers had installed a relatively large number of shallow tubewells even when they had access to deep tubewells and kulo. The use in Madhaulia, in comparison was very low. One of the main reasons to compare the use of shallow tubewell in other case studies in this chapter is to highlight the way the same technology is used in different ways in different sites. Shallow tubewells had become more or less an 'optional' tool in the other two sites while it emerged as the primary option for irrigation amongst farmers in Mahuwari.

The farmers in Tikuligarh VDC, who had access to kulo water for both paddy and wheat, and whose land fell on the tailend of the deep tubewell systems, installed shallow tubewells. The shallow tubewells were used to mitigate scarcity in the kulos as well as to irrigate vegetables that were either grown for personal consumption or sometimes grown in a larger scale and sold in the market. Farmers who did not have access to kulos (like those in Tikuligarh gaon) used the shallow tubewell to bargain their share of payment for flat rate of electricity for deep tubewells and also used to irrigate vegetables. Some farmers in Durganagar, who had

FIGURE 6.1: The defunct kulo and shallow tubewells with pumpsets in Mahuwari village

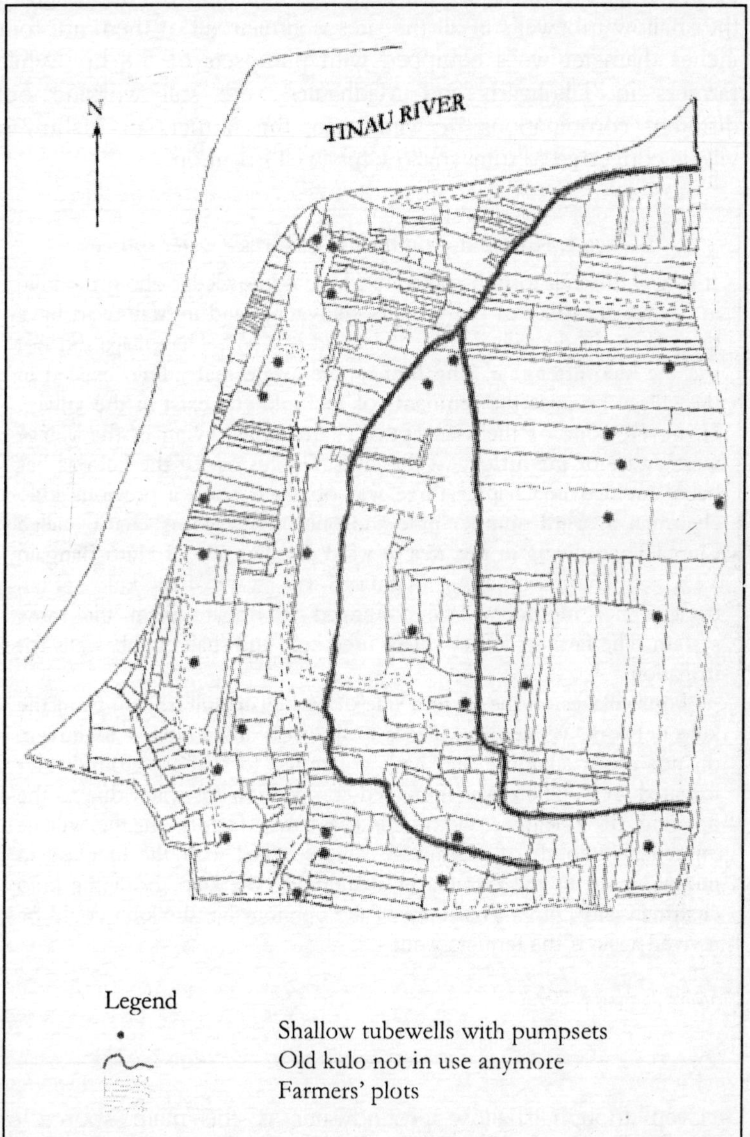

Note: The location of tubewells without pumpset is not shown in the figure.
Source : Field Survey, 2002/04

installed shallow tubewells, sold the pumpsets in 1999 after getting access to cheaper deep groundwater. The remaining pumpsets also provided an insurance against erratic power supply. The design of the shallow tubewells in all the sites is similar: all of them are four inches diameter wells equipped with pumpsets of 5-8 hp. While farmers in Tikuligarh and Madhaulia were still working out different combinations of water use, the farmers in Mahuwari village converted to total shallow tubewell irrigation.

BOX 6.1: Knowledge on the use of surface water sources

It is very difficult to find someone who has knowledge about the kulo in Mahuwari. Most of the farmers cultivating land in Mahuwari have been living in the village after the flood of 1965. The village did not use the kulo after that. The farmers are aware that a kulo existed in the village because the remnants of the kulo still exist in the village. However, none of them except a Tharu family living in the village were aware of the history of the kulo. The history of the kulo as has been outlined in Chapter three was obtained from a previous kulo chairman in ward number nine and one of the oldest Tharu called Durga Tharu living in the area in ward number one of Hatti Bangain VDC. A Tharu woman, in Mahuwari was aware of the kulo system because her mother village continued to irrigate from the same system. The farmers in her village used kulo and shallow tubewells for irrigation.

Four villages on the eastern side of Mahuwari still irrigate from the kulo network. While five villages to the west and south of Mahuwari do not use it anymore and have converted to shallow groundwater irrigated areas. There was increased competition for water due to the opening of a leather factory and a brick factory in the village immediately south of Mahuwari. This coupled with the increase in migration led to the farmers in abandoning the kulo. Both the kulo chairman and Durga Tharu are of the opinion that the kulo could be revived again if the farmers want.

Source: Fieldwork, 2002.

Even though shallow groundwater is the main source of irrigation water in Mahuwari village, there was no specific state-sponsored shallow tubewell irrigation programme for this village or for the VDC. There were also no community-initiated development works in this village before the end of the 1990s. The introduction of development programmes adopting the participatory approach started later in this village. After implementation of the participatory district development programme (PDDP) under the Local Governance Programme of the government, all development works were to be carried out through 'communities'. It was then, that the farmers in Mahuwari got a tubewell through this programme in 2001. All the farmers inside Tikuligarh and Madhaulia VDC, who had installed shallow tubewells, had obtained it through the ADBN loans. However, majority of the farmers in Mahuwari had installed shallow tubewells privately making use of their own social networks.

There were thirty-seven pumpsets and twice this number of tubewells in the village in 2004. Only eleven farmers had obtained shallow tubewells (pumpset and tubewell) through the bank. All eleven shallow tubewells in Mahuwari were installed before 1992. The rest of the tubewells and pumpsets were obtained by the farmers privately. Even though the ADBN was still carrying out its installations beyond 1992, none of the farmers in Mahuwari installed shallow tubewells through this means.

Records show that the farmers in Tikuligarh and Madhaulia VDC installed shallow tubewells throughout this period through bank loans, up to 1995. Well certification records show that at least 43 shallow tubewells were installed in Tikuligarh VDC from the period 1990 to 1995 (ADBN, 1997). This period coincides with the period before and after transfer process of deep tubewells. Field verification in Madhaulia showed that tubewells were installed through other means then the ADBN[1].

Farmers in Tikuligarh and Madhaulia VDCs made use of government sponsored programmes to access shallow tubewells while farmers in Mahuwari on the other hand, used different social networks that they had developed, for accessing tubewells and pumpsets (see Box 6.3 page 187). This network had not penetrated into the other two VDCs, even though distance between them is quite small. The southern part of Tikuligarh is less then a few kilometres from this village, separated by some villages in Chihiliya VDC. The travel distance between Mahuwari and other two sites via the Bhairahawa-Butwal highway and Lumbini road is around twelve kilometres. The process of installation of tubewells and purchase of pumpsets increased rapidly in the village after 1997. The farmers in this village are more involved in working out strategies to gain control over shallow groundwater and moving further away from surface irrigation.

Farmers' choices in shallow groundwater in this village are driven by various factors. Absentee landlordism provided an opportunity to the smallholders and gave the landless access to land for cultivation. The prospect of profitable agriculture has led to a contest for shallow tubewells amongst the farmers. This further led to the emergence of both individualised as well as collective forms of irrigation around groundwater.

In order to understand the move of the farmers in this village towards groundwater irrigation and further neglect of the surface sources, it is necessary to understand the background of these farmers, their objectives in agriculture and the reasons for their 'non-involvement' in surface irrigation.

Mahuwari: Socio-economic Reality

The village of Mahuwari is made up of two of the seventeen settlements that make up Hatti Bangain VDC. The two settlements are called Mahuwari and Kutti Tola, collectively referred to as Mahuwari. The settlements on the eastern part of the VDC are connected to Siddharthanagar municipality. Even though the eastern area is an urban fringe, the rest of the VDC is agricultural area. Bhairahawa airport is also in this VDC. Mahuwari is connected to Lumbini highway through a gravel road that passes through the village of Bangain. The office of Hatti Bangain VDC is situated in Bangain and lies on the way to Mahuwari village.

Most of the houses in Mahuwari are temporary dwellings. They are mostly huts made up of brushwood, stones and mud. It is a sharp contrast to the villages in the other case study areas, where the majority of the houses were permanent dwellings. A large part of Mahuwari (85 percent of it) belongs to absentee landowners. The rest of the land inside the village is owned by 67 households. There were in total 82 households living in this village in 2004, fifteen out of which were landless. Records show that the average landholding in Kutti Tola and Mahuwari Tola is 1.25 hectares and 0.80 hectares respectively (PDDP, 2000). But this average gives a very general picture of the reality. The land tenure is highly inegalitarian because a large percentage of land is held by a few people. The land distribution is skewed because sixty-four households own only 15 percent of the total land inside the village. The real average land holding for the rest of the farmers comes below 0.17 hectares. This in the Terai context belongs to the 'category' of 'very small farmers'. According to a recent survey, only 28 percent of the farmers in Mahuwari and less then 22 percent in Kutti tola relied on agricultural production from their own plots for one year (PDDP, 2000 survey on food self sufficiency).

The population in the village in 2004 was 643(337 men and 306 women). This was the number registered at the VDC. The entire population is of Bhojpuri speaking community, of north Indian origin. They fall under the North Indian caste system: Yadav, Lodha, Kewat, Kurmi, Teli, Mallah, Brahmins, Bhar and Chai. There are only two Tharu households. The social and cultural links of the farmers in this village with the villages in neighbouring Indian state of Uttar Pradesh are very strong. The border with India is only around five kilometres from the village. Most of the farmers migrated here at different phases, starting from the 1960s. Few of them migrated from other villages inside Nepal, but most women who married into the community came from India. All farmers who own some land in the village have citizenship papers. Other landless and the extended family members who have been working in agriculture for many years do not. The migration process was ongoing because some farmers had recently come to take up farming contracts. Child marriage[2] is prevalent amongst the lower castes and the dowry system is also still carried on. Illiteracy is high in the village, with only ten men who have completed high school.

The amount of labour for agriculture is higher here than in the other two sites, because everyone in the household from the age of twelve is fully involved in agriculture. Except for the village-level politicians and few members of the society who worked in the nearby sugar factory or in the VDC, most of the farmers did not have any linkages with the political and administrative bodies in the area. The landless and the very small landholders worked both in agriculture as well as in the construction sector in Bhairahawa. It was very difficult to find men in the village in the daytime. All men who worked as construction labourers performed their agricultural tasks in the early hours of the morning and the evenings, adjusting their schedules depending on the season. The rest of the family members carried out their tasks at other times during the day.

Other types of labour migration are going on in the village as well. The landless are mostly migrants from the Indian state of

Uttar Pradesh who have come to work in agriculture. They came through contacts with their kin living in the village. Thirty men worked as contract farmers for absentee landlords in 2004. Some worked both in sharecropping as well as in contract farming. All landless were involved in contract farming. Eleven men worked outside the country. Out of this, three of them worked as labourers in Saudi Arabia, three in Malaysia and five worked in Punjab in India. Their families managed the plots while they worked abroad. In most cases, the parents or brothers took the responsibility for overseeing farming. Ten people from the village had regular jobs in the now defunct sugar mill and two were working in the VDC office. Those who had jobs in the sugar factory were trying to retain their jobs by registering their attendance every day at the factory, with the hope that it would open again and that they would receive their salaries.

In terms of party politics, the entire village supported the Nepal Sadbhavana[3] party. Some of the larger politicians in the area lived in neighbouring villages. The VDC chairman in 2002[4] was from the Sadbhavana Party and lived in the village next to Mahuwari.

Changes in Cropping Pattern

The Mahendra Sugar mill opened in Rupandehi in 1963 and is a few kilometres from Mahuwari village. Most of the villages in and around the factory cultivated sugarcane. The farmers in Mahuwari likewise, supplied sugarcane to this factory. Besides cultivating and selling sugarcane, many people in the area were also employed in the factory. The people in the village however, experienced a setback after 1996, when the factory closed down due to internal problems. They no longer had a market for their produce.

This incident set off a new process whereby the farmers started converting their cropping pattern. They switched over to a paddy-wheat-mustard cycle and grew vegetables as cash crop. In 1998, the District Agriculture Office (DAO) introduced banana cultivation. The first plots of bananas were grown by a few enterprising farmers who set up a banana growers' group. At the time of the field survey in 2001/02, four farmers were still growing sugarcane. However, all farmers completely stopped sugarcane cultivation from 2003. Only those farmers who were able to make arrangements for transporting sugarcane on their own continued to do so. For this purpose, they hired tractors and transported the produce to neighbouring district of Nawalparasi. The farmers could no longer take the risk of moving the produce on their own, because of the escalating conflict situation in the country. Many farmers, instead, opted for banana cultivation since 2003. In 2004, they followed a diversified cropping pattern consisting of paddy, wheat, mustard, vegetables and bananas. Vegetables were sold in Bhairahawa and in other weekly farmers' markets. The farmers were directly involved in marketing of vegetables. Bananas were sold to wholesale dealers who came into the village to buy the produce right from the farmer's field. This saved the farmers from the risks involved if they had to transport the produce on their own.

PICTURE 6.1: Irrigation of banana plants using a shallow tubewell

The switch from sugarcane to vegetables and then to bananas has been possible in Mahuwari thanks to opportunities that provided access to land to the landless and the small holders. It has been mentioned above that 85 percent of the total land in this village is owned by three people who do not live in the village. Each landowner had a person in the village, who acted as a contact person or manager (locally called the sirbaar) for them. This person was given total responsibility for all decision-making regarding choice of crops as well as sharecroppers.

When sugarcane was grown in the village, smallholders and landless were hired as wage labourers. The manager was responsible for selling sugarcane to the factory. The money was divided between the landowner and the managers. Sharecropping (adhiya) followed wage labour after the closing down of the sugar factory. Sharecropping was done for paddy-wheat and was not profitable. Sharing between the landowner and the farmers was done on fifty-fifty basis. The start of banana cultivation led to the emergence of a new form of arrangement locally called *hunda*. This is a form of contract farming, in which the cultivator pays a certain amount of money to the landowner per year. The managers were very active in working out contract farming. Intervention in banana cultivation provided them with an opportunity for a new cash crop. The same managers first tried banana cultivation and later started giving out contracts after realizing the benefits. Contract farming gave an opportunity for the landless migrants to make a living and for the smallholders to increase the size of landholding. Some farmers came in to take a contract from neighbouring villages in India through their kin who had been to the villages earlier.

A total of thirty households had taken up contract farming in the fields that belonged to absentee landowners. The rate of the contract was fixed as NRs. 17700 per hectare for banana cultivation and NRs. 15700 for those farmers who grow paddy, wheat and a third crop in 2004. Sharecropping is also followed in case of the second cropping pattern. Farmers are required to pay cash directly to the managers. Some plots are owned by people who live in the district. In that case, the farmers and the landowners are in direct contact with each other.

Farmers tried to follow a combination of contract farming and sharecropping because of the risks involved in contract farming. The person who takes the contract is required to pay a lump sum under any circumstance. In 2003, a hail storm, affected the crops. The farmers were left with less margin than they had anticipated. After observing the experience of the pioneer banana cultivators in the village, more farmers ventured into banana cultivation (see Box 6.2). By the spring of 2004, more than 30 percent of the village land was covered with banana. The farmers who took up contracts in this village, also took up similar contracts in other villages in wards one and nine. After observing Mahuwari, farmers in the neighbouring village across the Tinau had also started cultivation of

BOX 6.2 The economics of banana cultivation

The variety of banana cultivated by farmers in Mahuwari village is Harichaal. This variety is planted in the month of May. It is harvested sixteen months later. The number of plants planted per hectare of land is around 3700.

Farmers in Mahuwari reported that they applied an average of 1 kg of fertiliser per banana plant. This is the maximum dose that the farmers apply. This includes 400 gms of urea, 400 gms of DAP and 200 gms of potash.

In 2002, the cost of urea per quintal was NRs. 1600. The cost of DAP was NRs. 2000 while potash was NRs. 1200. Total cost for fertilizer for bananas planted in a hectare of land was therefore NRs. 58000.

In terms of labour: one person digs an average of one 0.034 ha of land. Each person has to dig at least six times to pull out the new plants and replant them. Thirty labourers work on one hectare of land. Labourers include both hired labourers as well as the labour contributed by family members. Labour cost was NRs. 100 per month. Total cost for labour was NRs. 27000.

An average of 8-10 dozens of banana grows per banana plant. The market price of banana was NRs. 8 per dozen in 2002. Even if only 3200 banana plants survived, the total money obtained amounts to NRs. 204800.

Farmers spent an average of NRs. 10000 on irrigation (when only their directs costs in shallow groundwater irrigation are calculated), the net profit of the farmer per hectare was NRs. 109800. In such a situation, farmers who took up contract farming paid NRs. 17700 to the landowner and kept the margin. A farmer who followed sharecropping would have to pay half of the net produce, leaving them with much less than when they take up contract farming.

The marketing of banana is done through the group as well as individually by the farmer. The whole sale dealers visit the village. The rate of the banana is fixed in the village. If the banana is of higher quality it sold at a rate of NRs. 10 per dozen and if it is of inferior quality the dealers bought it at NRs. 8 per dozen.

Farmers in Mahuwari reported that they use more fertiliser than needed because they believe more fertiliser application improves the quality of the crop. They do so because of fear of loosing the crop. They irrigate the crop seven to eight times. In the dry season the farmers do not to take a risk of a dry spell.

Source: Field Survey, 2002

banana by 2003.

Any person who took up a contract was responsible for the cost of all inputs. Therefore, all the farmers, from landowners to landless, who had taken contracts were engaged in finding out strategies to reduce the cost of inputs. The farmers stated that they did not compromise on the volume of water or the quantity of fertiliser used for banana. When they worked as wage labourers in sugarcane cultivation, they did not have to take the responsibility for the inputs. Those were the responsibility of the landowners or the manager. However, once farmers took up a contract, they had to ensure adequate and reliable water supply on their own. Hence, they tried to drill more tubewells and find out ways of getting cheap pumpsets. All this involved cash. For this, the farmers made use of loans through village-based savings groups. In the formal banking system, a person has to keep land as collateral to obtain a loan. In the rural banking system anyone can join a group and take loans. Different types of group exist in the village. There were three groups initiated by the PDDP, one group through the Agriculture Development Branch and one formed by the Women Development Branch of the government. The group formed by the Agriculture Development Office was the banana growers group. There were fifteen members with a monthly savings scheme of NRs. 20. The group met once a month to discuss the problems in banana cultivation, collect savings and make arrangements with the agriculture office for an agricultural technician.

Strategies to gain control over shallow groundwater

Groundwater use in Mahuwari had started with the use of privately owned shallow tubewells and pumpsets and always remained individualistic. The trend to install privately owned tubewells and obtain a pumpset increased through the years.

Subsidies for group shallow tubewells through the bank, however, existed in the 1980s. Except for one farmer, none of the farmers in Mahuwari had accessed the loan for 'community' shallow tubewells. Even though it was accessed by making use of the subsidy for community shallow tubewells, it is a case of what Koirala has called as 'fake community shallow tubewell' in his study on different types of community shallow tubewells in twenty districts in the Terai (Koirala 1998:29). The loan was accessed by the farmer by officially fulfilling the 'criteria' set up for community shallow tubewells. A request for a group tubewell required a minimum of three farmers with a contiguous plot of four hectares. This man manipulated by combining his land and portions from the land owned by an absentee landlord in order to fit into the criteria and access high level of subsidy. Most community tubewells ultimately became individually owned after one person paid off the loan. It was only in 2001 that the farmers initiated a group pumpset in the village.

There were twenty-seven pumpsets owned by sixty-seven households in Mahuwari in 2004. These pumpsets were shared by all households. The farmers had to use the pumpsets not just for sixty-seven landholdings but for a larger number of plots. There is a high degree of land fragmentation, due to both inheritance as well as the system of giving out land contracts. Moreover, each farmer owns small plots in more then one place. Land scatter is one of the basic structural elements that affect the emergence of different practices for water use. Farmers are not able to irrigate all their plots, even if they have a private pumpset.

The farmers worked out different strategies to access water. The common practices were: renting pumpsets, lending tubewells, buying and selling water and exchanging water. A farmer who owned a pumpset, first completed his own irrigation. If he had more then one plot he had to move the pumpset to the next plot. However, this action was subject to their having right of way through their neighbour's fields. When other farmers had a standing crop, it was difficult to move the pumpset from one place to another. In that case they rented in pumpset from the neighbour. Many farmers tried to install tubewells in their several plots. This had become easier because of the availability of local drilling mechanics in the area (see Box 6.3). When they did not have a tubewell, they either bought water from the neighbour or 'lent in' the tubewell. A practice had evolved in Mahuwari for letting each other use tubewells free of cost. Originally, a charge of NRs. 10 per hour was set for using a tubewell. However, after the number of tubewells started increasing in the village, the farmers let each other use tubewells free of charge. They felt it was their *bhyavahaar* or etiquette to do so. Sometimes farmers had the same neighbour (brothers) in more than one plot. This was common when the land had been inherited from the parents.

Groundwater was also traded in different ways. A tubewell owner sold water after completing their irrigation. The rate in use

BOX 6.3: Emergence of the shallow tubewell enterprise

Ram Samoj Kuhar is the main person who drills almost all the shallow tubewells in Mahuwari. He says he is not as happy as he was a few years back. According to him, 'Rupandehi District is nearly punched out because there are too many wells being drilled'. There are also many drilling mechanics. Ram Samoj has been drilling wells in Mahuwari and the neighbouring village for fifteen years. He lives in the neighbouring village. According to him, in the 'good old days of the ADB/N involvement in shallow groundwater', he was one of the few drilling mechanics to be pre-qualified by the bank. 'Previously, I used to drill twenty-five tubewells in a year, but these days, competition is getting more tough; I get an average of seven tubewells in a year. This year I drilled seven shallow tubewells and an artesian on the Nepalese side and two across the border in India'. Ram Samoj has a team of labourers to drill a well and he is the head *mistri* or mechanic.

Ram Samoj is one of the nine drilling mechanics (locally called mistri) who are involved in drilling shallow tubewells in Mahuwari and the neighbouring villages. He is the only one who was trained by the bank[1] and legally qualified as a mechanic for drilling shallow tubewells. However, other mechanics have come up as the skill has been handed down through experience from one to another. After the bank withdrew from the technical component of shallow tubewell installations, the drilling mechanics had to find out their own means by which they could earn a living without agency support. The pumpset dealers too had to find out avenues to convince the farmers to use their brand of pumpsets.

The mechanics like Ram Samoj are very important actors in shallow groundwater irrigation in the area. They serve two functions. One: they actually arrange the drilling of shallow tubewells and flowing artesians. Two: they are the main person involved in promotion of different pumpsets at the local level. Their authority and influence has risen after the withdrawal of the bank monopoly on shallow groundwater irrigation. In fact they are now the main designers and promoters of shallow tubewells. The fact that they live in the villages have made them the most powerful extension agents

Source: Fieldwork, 2002/2004

in 2004 was NRs. 100 per hour. The farmer who bought water could bring in own fuel and use the pumpset in the tubewell of the person who owned it. This was done in case of close relatives or with other persons with whom the farmer shared good relations. In that situation, the person borrowing the pumpset was trusted to take part in repair in case of breakdown. However this was not practiced much. Farmers, who could not irrigate a certain plot because of difficulty to carry their pumpsets, let their neighbour use their tubewell if they had one. They then irrigated through the same pumpset. A payment was made if the two did not have plots next to each other in another location. When they did so, and if the first farmer had a pumpset, he let the second farmer use it. In this way they exchanged water. Farmers installed more number of tubewells in order to make transactions in pumpset renting and also transportation of water from one plot to another easier and less complicated.

The rate of renting, as has been mentioned above, was fixed at NRs. 100 per hour. This rate was increased to NRs. 120 per hour in 2003, after the increase in the price of diesel during the Iraq war. In 2004, the farmers lowered the rate again to NRs. 100. This was the rate throughout the district. The rate of NRs. 100 per hour was fixed based on the average consumption of diesel by the standard diesel engines used. The average size of the diesel engines used was 5-7 hp. The fuel consumption rate in these engines is around 1.15 litres per hour. It is taken as 1.50 when the engine is not of good standard. The price of diesel in 2004 was NRs. 31 per litre. The flat rate of NRs. 100 per hour was made up by adding the value of fuel consumption per hour and the rate for the rental of the pumpset only. The latter was NRs. 50 per hour. None of the farmers paid more than the prevailing rate and they preferred cash payment over other forms of payment. The other common practice in Mahuwari was river water pumping. Farmers who had land next to Tinau River pumped water directly from it.

Nine farmers were selected in order to understand how the farmers in this village irrigated and the strategies they employed and the networks they made use of in the process of gaining control over groundwater. One of the main objectives of the farmers is to minimize the cost of irrigation. I therefore present the analysis in tables (Table 6.1, 6.2) that show the costs of operation of shallow tubewells for selected farmers and the ways they try to reduce their costs. Besides the costs of operation, several things can

TABLE 6.1: Individual Strategies of Farmers to Gain Control over Shallow Groundwater (1)

	Farmers								
	1	2	3	4	5	6	7	8	9
Land holding (ha)	1.28	1.69	1.01	0.42	0.68	3.42	0.34	0.51	2.03
Engine capacity	5	7	8	8	7	6.5	7	7	8
Pump set cost (NRs.)	7000	19000	16000	4500	7000	10000	22100		8000
Boring cost (NRs.)	8000	4000	10000	3000	10000	5000			7500
Total cost (NRs.)	22000	23000	26000	7500	17000	15000	22100		22500
Fuel consumption (lit/hr)	1.15	1.5	1.15	1.25	1.5	1.25	1.5		1.5
Hours used in irrigation & threshing	82	107	26	17	38	100	35	95	107
Lubricant cost (NRs.)	75	60	60	50	50	60	40		60
Fuel & lubricant (NRs.)	180	250	99	83	132	200	113		250
Repair & Maintenance (NRs.)	1600	1600	1600	1600	1600	800	1600		1600
Depreciation	2200	2300	2600	750	1700	1500	2210		2250
Interest @18 % per annum	3960	4140							

Total operating cost (NRs.)	7940	8290	4298	2433	3432	2500	3922	4750	4100
Cost per hour	96	77	165	147	90	25	112	50	38
Discharge (m3/sec)	0.01	0.02	0.04	0.03	0.02	0.01	0.03	0.03	0.02
Volume used (m3)	2952	7701	3744	1782	2736	3600	3780	10260	7704
Cost per cubic meter (NRs./m3)	2.69	1.08	1.15	1.37	1.25	0.69	1.04	0.04	0.53
Total direct costs (NRs.)	1780	1850	1698	1683	1732	1000	1712	0	1850
Direct cost per hour	21	17	65	102	45	10	48	0	17
Total indirect cost (NRs.)	6160	6440	2600	750	1700	1500	2210		2250

Source: Field work, 2002

TABLE 6.2: Individual Strategies of Farmers to Gain Control over Shallow Groundwater (2)

	1	2	3	4	5	6	7	8	9
Actions taken by the farmers									
Buy water (hrs)	5	20	15	10					10
Cost buy (NRs.)	500	2000	1500	1000					1000
Sell water (NRs.)	50	10							
Earns	5000	1000							
Farmers perceived cost (NRs.)									
Fuel	2451	4667	1155	2342	1396	3062	1286	4750	3932
Lubricant	75	60	60	50	50	100	40		60
Repair & Maintenance	1600	1600	1600	1600	1600	800	1600		1600
Total cost acc. To farmer (NRs.)	4126	6327	2815	3992	3046	3962	2926	4750	5592
Costs after taking actions	3440	9290	5798	3433				4750	5100

Source: Field work, 2002

be discussed by means of this table: it shows the type of farmers, the technologies in use, from where they obtain the technology, the networks they make use of and the types of institutions that are coming up around shallow groundwater. All farmers in this table are landowners.

The second table is to be read in line with the first table as it is the continuation of the details of the same farmers in the same order. The first two farmers installed their shallow tubewells and pumping sets through ADBN (Narayani Kirloskar) loan. The third had drilled a tubewell at own cost. He tried to get a pumpset from the bank but could not, because the bank did not provide subsidy for pumpset only. He bought a used pumpset (Narayani Kirloskar brand) from a farmer in the neighbouring village. The first farmer installed a tubewell again at his own cost to irrigate banana.

The fourth farmer owns only 0.42 hectare of land. He drilled a tubewell by contacting the local drilling mechanic. He did not go through any agency. This farmer has relatives across the border. He made contacts through these relatives and got a very cheap Ajit brand pumpset. His contact got it for him from a Harijan family in India (the highly subsidized pumpset). Similarly the fifth, sixth and ninth farmers also got second-hand pumpsets from India. The seventh farmer did not buy a pumpset by himself. He got it as part of ancestral property from this father, who got it as loan from the bank. When the property was divided, the tubewell went to his brother and he got the pumpset.

The range of values for the pumpsets is different because it depends on the state of the pumps and the brand name. Even though the sixth farmer bought a second-hand pumpset, he made sure that he got a good one. He spent more money than the others who got it second-hand from India. The eighth farmer does not own a pumpset. He rents in a pumpset and irrigates from the river.

Farmers number five and six are owner cultivators. The rest do cultivate their own fields but also take up sharecropping and contract. All farmers cultivate part of the field with paddy, wheat and mustard, and the rest with banana except for farmer number seven. He does not grow paddy wheat but only bananas and vegetables. He tries to maximise his income by cash cropping. He works as a construction labourer too.

Table 6.1 shows that the farmer who invests least in groundwater technology is the fourth farmer. Of course the eighth one does not invest at all. He is a member of the shallow tubewell group and rents in the pumpset at NRs. 50 per hour. But he is also the one who uses the largest volume of water for irrigation. The eighth farmer uses a high volume of water because he grows bananas. The cost per cubic meter of water is also lowest for this farmer. The table shows that even small farmers who do not possess a pumpset are able to go for diversified cropping. They take up banana contracts and make up for the costs involved.

This table also illustrates the different relationships that the farmers are involved in. Even those farmers who own a pumpset or tubewell have to buy water. This is because they have plots in different locations and their pumpsets cannot be carried to each plot every time. At the same time they sell water.

Farmers do not calculate the indirect costs. Their perceived cost is based on the operation of the pumpsets (Table 6.2). However the real costs of operation are high (table 1: sum of direct and indirect costs). If the eighth farmer would be buying directly from other pumpset owners, he would have to pay a rate of NRs. 100 per hour. Now he pays half that price by renting in the group shallow tubewell.

The size of the shallow tubewells is uniform. They are all of four inches diameter tubewell with a pumpset from 6-8 hp capacity. This is the same irrespective of the fact whether they are individually owned private shallow tubewells or group tubewells. There is redundancy in the technology. For example, farmer number three uses the pumpset in his 1.01 hectare land for only 26 hours. For a paddy wheat cropping pattern in this size of plot, this is sufficient. He does not sell water but, instead, rents in a pumpset for one of his plots.

This redundancy is caused by several factors: one: the mismatch between decreasing land holding and the capacity of the technology. Two: the personal ambition of the farmers to have total control over water, which induces them to buy their own pumpset. Renting in and lending out involve social interactions, which are hard to maintain at all times. One of the main causes of tension is that fact that farmers have to transport their pumpsets from one plot to another. They have to do it in such a way that they do not spoil the crops of the others on the way. There are no separate paths or roads to the plots.

Most farmers in Mahuwari are involved in installing shallow tubewells on their own. This has been possible due to the emergence of a network of actors who are involved in the shallow tubewell and pumpset business (See Box 6.3 page 187). This network consists of the drilling mechanic, the pumpset dealers, and the networks that the farmers have developed through the mechanic or through their own kinship relations in India.

PICTURE 6.2: Drilling for a flowing artesian well

Group strategies for control over groundwater

The farmers in Mahuwari obtained a group shallow tubewell in 2001 through the PDDP programme for implementing priority productive investments. This came as a respite to those farmers who had to rent in a tubewell or buy water from their neighbours. The shallow tubewell came through the savings group initiated through the VDC, by the PDDP. Mahuwari Samudayik Sanstha (Mahuwari Community Organisation) was a savings group with forty members. All member farmers in this group were eligible to take loans from the savings account. They could pay back the loan at an interest of 18 percent. An office for the social mobiliser of the programme was situated in the VDC. As per the programme she was responsible for initiating community organisations throughout the VDC. The farmers in the savings group expressed the need for a pumpset during interaction between the staff and farmers. Different groups in the VDC requested different technologies through these processes. For example, the farmers in Mahuwari requested a pumpset, but a group in neighbouring ward number one, requested drinking water. The programme helped install a flowing artesian well next to the highway. This well served as a public drinking well for the community as well as others who passed by the highway. In the case study in Bihuli, the farmers mobilised money through the same programme in that VDC for gabion boxes for headworks in Lausi Khola.

One of the main persons besides the staff of the programme, who took the initiative to convince the farmers about the savings was Sukhram. He also worked as a clerk in Hatti Bangain VDC. Even though his position in the VDC administration was a minor one, he was the main person in Mahuwari who was in constant contact with development programmes that came in through the VDC as well as other contacts and information. He was also the main source of contact for the villagers for any official work that had to be done through the VDC. Besides this, he was manager for the land of one of the absentee land lords and had given contracts to four farmers who cultivated bananas. Sukhram was made the chairman of the savings group. All farmers in the savings group maintain pass books where their accounts are managed individually. Sukhram was also in charge of maintaining these in his capacity as the VDC clerk.

The total cost for the shallow tubewell was estimated at NRs. 43000. The group members had to contribute 10 percent as labour contribution during installation. They were further required to save five percent on the total cost. The tubewell had to be installed somewhere in the village. The farmers agreed to install it in the chairman's plot as he had to take the responsibility for the pumpset and coordinate the lending process as well. Each group member had to contact the chairman to rent the pumpset. Not all forty members of the group actually used the pumpset. The group was not formed as an irrigation group. Eighteen members were using it in 2004. Some of the members already had their own pumpset. The cost of renting the pumpset was NRs. 50 per hour in 2004. This is half the price of renting private pumpsets. The group saved NRs. 10 from each hour of pumpset use and put in the savings fund; rest was spent on for diesel and maintenance. By 2004, many complaints had surfaced regarding the use of the pumpset. Most of the complaints centred on diesel consumption and farmers not maintaining the rental hours. The group later revised the rules. In 2004, the rental charge was fixed at NRs. 20 per hour. This was the charge for renting in the pumpset only. Farmers had to put in their own diesel from that time.

One of the main causes of conflicts that emerged in the spring of 2004 was the difference of interests between the banana growers and non-banana growing members of the group. The banana growers had the tendency to over irrigate and not take any risk of drought. This made them hold on to the pumpsets for longer hours.

Box 6.4 Setting up the pumpset

It takes a farm family around 25 to 30 minutes to mount the pumpset before they start to irrigate. The time to do so depends on the state of the pumpset. It also depends on how efficiently they are able to make the system. Many farmers in Mahuwari do not leave the pumpsets in the field for fear of being stolen or tampered with. The base of the pumpset is mounted on wheels. This enables them to transport it to the tubewell site. The farmers push in the wheels into the soft soil in order to prevent the pumpset from moving while in operation. They use few barks, stones and wooden planks to make it more compact. None of the farmers in Mahuwari had made a permanent base for mounting the pumpsets.

The next operation is the priming of the pump. Before that, the farmers try to make the system airtight. The men and women mix clay with water and make a plaster. This soil is plastered in different areas where there is a chance of air leakage. A hand pump is mounted for priming purposes. One person plasters rubber sole slippers with the mud plaster and sticks this to the mouth of the hand pump. Another person adds water from the bucket to the opening next to the piston of the hand pump, while the other one goes on pumping water by moving the handle of the pump, till the water starts coming up the pipe. All this time, the person holding the slipper presses it very tight. When water starts gushing out of the shallow well, the force throws away the slipper and it is that very time when they start the diesel engine.

There were complaints from the others, that the banana growers rotated the pumpsets amongst themselves. There were also complaints against the chairman that he gave the pumpset to one of the farmers whom he had given a banana cultivation contract. I saw during the fieldwork that the banana farmers did not wait for the rains. Sometimes they irrigated early in the morning, feeling very unsure of the rains. The same afternoon we experienced heavy rainfall. The farmers planted bananas in the last week of May or in the first week of June. The plant takes more than fifteen months to mature and is harvested after that period. It was irrigated at an average of ten times in the whole cycle. It was irrigated in the dry season starting April to beginning of June and then again in October to January when there was no rain. The other farmers needed water at the same time in May-June for paddy nursery and then for paddy transplanting. Paddy is transplanted on (around) 29 of June, wheat is grown on (around) 25th November for the majority of the area (paddy for 2002 and wheat 2001).

This was the first time the farmers had come together collectively as a group. According to the chairman, managing the group was a problem because of the size of the group. Moreover, he was busy in the VDC and was not able to monitor the activities of the farmers. There were also discussions on splitting the group according to socio-economic status or cropping pattern (paddy-wheat and banana) whichever would be suitable. This would make it easier to make rules and enforce them. There was nobody to monitor how long each farmer irrigated from the pumpset after it was rented. Most farmers used the pumpset for a longer time than they paid for. They justified their action under the argument that it took them an hour just to move the pumpset to their plots, fix it up and then to operate it (see Box 6.4).

Conclusions

Possibilities of profitable agriculture was one of the major forces driving the farmers in Mahuwari towards gaining individual control over water and thus moving more towards shallow tubewell irrigation. Besides Mahuwari, five other former maujas had converted to shallow tubewell irrigation. Flood had destroyed part of the kulo in this village, but other villages too had left kulahi.

The farmers had been involved in cash-cropping for many years and tried to work out farming strategies through the years to incorporate other cash crops. This led them more and more towards groundwater irrigation. Farmers in this village were not interested to revive the kulo because they had already realised the importance of having total control over water. The case study shows how prospects of profitable farming induced the evolution of different farming strategies. This was also made possible thanks to the availability of shallow tubewells. This dynamic further induced the movement away from kulo irrigation.

This chapter also examined the different arrangements that the farmers have made for irrigation. The arrangements that arise for irrigation in areas irrigated by shallow groundwater are influenced by the physical properties of land such as land size and land fragmentation. The shallow tubewell technology both facilitates as well as constrains these arrangements because it can be used as two components: a fixed tubewell and a movable pumpset. The farmers work out different ways within these constraints. Different types of relationships have evolved between farmers even around this very 'individualistic' technology. They try to work out ways of irrigating in such a way that it does not create a hassle. The way they carry about their *'Bhyavahaar'* is one example. They have also developed ways of buying and exchanging water, renting in pumpsets and the practice of lending out tubewells. These activities also overlap with each other. The aquifers in this village are recharged by the Tinau River and the Ghagra Nala. The farmers have not experienced reduction in discharge from the tubewells and have thus developed this practice of lending out the tubewells. The emergence of the local 'drilling industry' has also facilitated the farmers to drill tubewells at reasonable costs.

Most of the farmers make use of their own networks for accessing irrigation technology. As a consequence of the individualised nature of the technology, the strategies of the farmers are also individualised. The processes and practices are directed towards finding different ways by means of which they can have more control over irrigation technology and groundwater. The farmers are thus involved in reducing cost of the technology by different means. Only few farmers who own land in the village made use of a government agency to obtain the tubewells. Most farmers used their own networks to access shallow tubewell technology.

This transformation in irrigation has been going on with minimal government involvement, in contrast with the situation in other VDCs. This does not mean that it is the preference of the farmers. There is active involvement when the intervention process caters to their needs as shown by the case of the emergence of the shallow tubewell group initiated through external intervention.

Notes

[1] A shallow tubewell was installed in Madhaulia in 2001, through the groundwater project in Butwal. The project is responsible for the technical component of shallow tubewell installations in the district after the withdrawal of the bank from this component. The project is implemented through eight groundwater field offices at different locations in the Terai, including one in Butwal in Rupandehi District. The programme includes installation of both shallow tubewells as well as deep tubewells through different approaches. The shallow tubewells of four inches diameter and covering at least 2.5 ha are provided subsidy of 60 percent (for groups) and 30 percent (for individual) subsidy. The remaining portion of the cost has to be borne by the farmers themselves or as a loan from ADBN or other financial institutions. The programme also includes a 'cluster approach'. If farmers living in areas covering 100-1000 ha demand shallow tubewells, the programme activities include other components like agriculture road, agriculture extension services research and development and rural electrification. Deep tubewells entail an 84 percent subsidy. The farmers have to bear the remaining cost, either in terms of cash or loan. The expected area to be irrigated by deep tubewells is 40 hectares. The construction includes a tubewell, pump house, electric motor and distribution system.

[2] The marriage takes place when the children are eight or nine years old. The girls stay with their parents until they come of age, when they reach puberty. A certain ceremony called the *gouna* is held. Unlike the Tharu and the hill migrants, the dowry system still exists in this community. The extent of dowry is higher when girls are married to boys from India.

[3] The Nepal Sadbhavana Party is one of several regional and ethnic parties established in Nepal after 1990. It was established as a forum for people of Indian descent, supports a democratic socialist society and promotes the interests of the Terai Region.

[4] All local-level bodies were dissolved in 2002 and the VDC was managed by the VDC secretary during the time of the fieldwork.

7

Conclusions

This study examined irrigation practices where farmers had the option of using more than one source of water for irrigation. The study was conducted in an area in the Terai of Nepal, where groundwater irrigation interventions predominate as a major focus of development policy. I examined how interventions in groundwater irrigation interplayed with existing management practices of other sources of water and how the management practices of groundwater itself have been influenced in the process. The interactions between different complexes of technologies and institutions in the larger hydrological, agro-ecological and political environment were studied in three different VDCs. Public interventions for deep tubewells had been the main focus of development in two of the sites, while shallow tubewell irrigation was the predominant intervention to transform irrigation in the third area. All the three sites had been irrigated previously by surface sources. I have also examined management of groundwater irrigation within conjunctive water use for irrigation in two of the study sites where farmers had access to deep tubewells. Collective practices around tubewells were also studied for all three sites. A review of key findings is presented in the first part of this chapter. I then discuss their implications with respect to current policy on groundwater irrigation and conjunctive water use. I also comment on the approach and methodology adopted for this study.

Incorporating Groundwater

Water use complexes: constraints and opportunities in securing water

All the three study sites lie within a complex water resource area in the Terai part of the Tinau River basin, whose rivers have been used for surface water (kulo) irrigation since settlement began in this region. The Tinau basin is also underlain by multiple aquifers of different depths, which were also identified as sources for irrigation water to augment agricultural production options. These aquifers of the basin were the focus of the largest national programme for deep tubewell development in the country - the BLGWIP - as well as public programmes for shallow tubewell development. This part of the river basin, like all other parts of the Nepal Terai, is subject to the monsoon rains, but its flow regime is shaped by its zone of origin. Unlike the rivers that originate from the Himalayas and snowmelt, rivers originating from lower hill ranges have higher variation in seasonal flow. They have regular challenges of concentrated flow in the monsoon, risks of flooding and sedimentation and low flow for the rest of the year. The Tinau has a more stable flow even in the dry season because it drains from a catchment in both the Mahabharat and Churia hill ranges, but floods are also common in the rainy season. These have shaped the demands on social organization for surface irrigation, especially for maintenance. One of the unique features of the hydro-geology of this area is also its connectivity, such that a substantial amount of water appears as jharan flow in lower reaches of the study sites, when areas in the upstream part irrigate. This study has shown that these diverse and often interconnected sources were actually approached as separate and distinct sources to be developed by intervention programmes. It has been farmers that have appreciated first their potential for conjunctive use that makes irrigation both more flexible and less costly.

The BLGWIP intended to open up parts of Rupandehi district for cultivation by supplying year-round water, with the programme starting up from the tailend of the existing kulo systems. It began in the mode of a large scale project under a supply driven mode as was typical of all irrigation development projects in the 1970s. Villages inside the project area were therefore supplied with deep tubewells that were capable of supplying year round sufficient water for irrigation. In later years, there were changes in the design of the tubewell as well as the development approach. It adopted a 'demand based approach' in the latter half of the 1980s and later into the 1990s. Changes in design were also incorporated so that the new tubewells were of lower capacity and thus incurred lower costs in operation than the older designs. The design modifications made the volume more divisible so that the farmers obtained the volume of water they required, and at a lower cost. The high costs of operation were one of the reasons why farmers in older tubewell areas were not willing to take over the tubewells. Shallow groundwater irrigation in the district started in the same manner as the rest of the districts in the Terai. It was used both as the main source of irrigation as well as a complementary source. Over time, there have been no changes in the design, but there have been changes in the development approach. Since 1997 there was more emphasis on group shallow tubewells.

The study shows that great diversity exists in the arrangements that the farmers made for irrigation. In the supposedly 'deep groundwater-irrigated' areas of Tikuligarh and Madhaulia VDC, farmers were organised around different complexes of water sources rather than only around deep groundwater. The farmers did revert back to using other sources of water after the transfer of the deep tubewells to them. However, analysis of the history of water use in these two areas showed that the farmers tried to fit the deep tubewells into their existing water complex from the initial period of intervention. Right from the beginning, the farmers regarded the deep tubewells as a complementary source of water. Several maujas in Tikuligarh still retained their water rights to the kulos, even when they had almost free groundwater supplied to them by the project. There were changes in relationships with the kulos later on. However, such changes were quite specific to particular villages. Whether a village could retain its kulo water right depended on the social relations inside the village. It also depended on where it was located along the surface irrigation system. Fear of flooding was also one of the factors driving the farmers to maintain rights to surface water and to maintain the drainage structures in the monsoon. Even though Mahuwari was irrigated only by individual shallow tubewells, the farmers were linked to each other through intricate relationships that were also shaped by the technology of the shallow tubewells and their commands areas options to share and rent irrigation water access through its elements, and land holding configurations.

Farmers using the older deep tubewells in the VDCs preferred deep tubewells for irrigating paddy and wheat, even if they were opposed to paying the flat rate of electricity. The high volume of water obtained from the tubewells in a short period of time was preferred for irrigation of major crops like paddy and wheat. The design of tubewells did constrain them in irrigating crops like vegetables that require frequent but lighter irrigation. Farmers used shallow tubewells for irrigating vegetables, as water delivery regimes of both kulos and deep tubewells (old design) were not suitable for this. They also used shallow tubewells as an insurance against electric power supply failure. Farmers preferred deep tubewells to shallow tubewells when their design was compatible to their needs, as in Durganagar.

Kulos provided farmers with advantages and flexibility, but also created insecurity. The advantage of using the kulo was that they had access to a better quality of water and could pay in labour rather than cash. The insecurity came in the difficulties in mobilising labour and cash demanded for as part of rights to use kulos, and the reliability and adequacy of water. Farmers in Tikuligarh still tried to maintain minimum rights to deep tubewells, even when they irrigated from surface sources. On the other hand, farmers in Madhaulia, who had more control over the Lausi kulo, depended less on deep tubewells.

Intervention in deep groundwater irrigation introduced completely new sets of technical, normative and organisational relations. First of all, the initiative set up a boundary between surface kulo networks and the project area in Tikuligarh and Madhaulia. A completely new process of (re) organisation around new water resources (groundwater) was started in these villages. The farmers who had already developed a certain form of relationships around kulos or jharans had to organise around deep groundwater. The infrastructure of the deep tubewell interplayed with the existing surface water irrigated landscape. As long as the project was supplying water, the differences and contestation around water sources were mitigated. However, after the handover process, the struggles were accentuated because each deep tubewell converted automatically into an irrigation unit struggling for its survival.

Different types of institutions were emerging for irrigation management even when groundwater was almost free. At the time of the fieldwork, farmers in Tikuligarh were organised around multiple complexes of water: kulos, jharans and shallow tubewells along with deep tubewells. Some villages in Madhaulia used deep groundwater and surface sources conjunctively, while others were using surface water as a major source of irrigation.

Conjunctive water use and conjunctive water management

The evolution of conjunctive use institutions was examined in detail in the case studies of Tikuligarh and Madhaulia. The BLGWIP project was not designed for conjunctive use: this was an outcome. Project documents and reports by those involved in the design and implementation do mention that farmers were using other sources of water besides only groundwater. A survey by Tahal (1992) noted that 44 percent of farmers reported that they used deep tubewells conjunctively with surface sources (in Olin, 1995: 216). However, this finding was not incorporated into the design process.

The farmers started optimising the use of deep groundwater immediately after the deep tubewells were handed over to them. This study showed that some forty percent of the supply for irrigation was covered by deep groundwater. The rest was covered by rainfall and other sources of water. The nature of the surface flows influenced when and how groundwater was used.

Farmers in Tikuligarh and Madhaulia mainly used deep tubewells with kulo and jharan water. Some also used shallow tubewells along with kulos and jharans. In Mahuwari, where only shallow tubewells were used, farmers also used water from flowing artesian wells or pumped water directly from the river. Chapter 3 showed the different complexes of water use including groundwater found across the case studies.

Co-ordinating and scheduling allocation between kulos and deep tubewells did not pose a problem for the farmers. Water supply to the villages from kulos was subject to a specific regime. Each village was informed about the time and date they would be allotted surface water by the larger irrigation system. Groundwater was used when needed, especially in the dry season and winter: at other times farmers demanded water when needed.

Shallow tubewell use or installation in a deep tubewell command area was found to be detrimental to the development of the deep tubewell as an institution for irrigation management. Farmers who owned shallow tubewells in deep tubewell command preferred to calculate the economics of use before making a choice between these two different groundwater sources. These farmers with shallow tubewells claimed that they irrigated only a small portion of their land from deep tubewells in order to reduce their share of payment of the flat rate of electricity. Even when they did so, they were still able to get a substantial volume of water because of the high discharge from the tubewell. Even farmers with access to two seasons of water supply from kulos had installed shallow tubewells for vegetable cultivation. Such farmers did not use deep tubewells but tried to maintain rights to it as a security by paying a minimum sum of money as per the demand charge. They used their shallow tubewells as an argument to legitimise their claim to a low payment, like those in Tikuligarh gaon. Farmers with no option besides the deep tubewells changed their winter cropping practices and opted for rainfed crops, because using deep groundwater for wheat was not economical.

The case studies also show the different ways in which farmers have been managing deep tubewells and kulos, or deep tubewells and jharan, at the same time, integrating them into their own governance spaces. Farmers in Supauli and Bihuli used deep tubewells with an open flow system with the kulo network; those in Durganagar used deep tubewells with pipe flow system with the jharan. Despite the supposed distinction in areas of kulos and BLGWIP, there were areas in the VDCs where farmers were able to access year round surface water. Some farmers in such areas in Tikuligarh VDC used both kulos and shallow tubewells, and did not need the DTW. The cases show the range of management strategies across the different conjunctive use complexes, which was shaped not only by the options of the system, but how farmers structured the wider governance of land, water and development options such as within the VDC. The way each complex was managed was unique in its own, but also bears similarities with other complexes because of the social as well as hydrological and technological dynamics.

Each deep tubewell committee set up its own plans, rules and regulations for managing the tubewells, and for raising income to cover its operation and maintenance. A comparison between Supauli and Bihuli shows how two villages with deep tubewells of the same generation (same design and subject to the same processes of reforms) have developed different management styles. The farmers in Supauli had to pay for the flat rate of electricity of the deep tubewell in order to claim their rights to surface irrigation. There was no such rule in Bihuli. This can be attributed to the fact that the conveyance system in Supauli for groundwater and surface water was common, while in Bihuli it was not. Groundwater is important source of irrigation for both Supauli and Bihuli. However, it was found to be a more serious business for the farmers in Supauli because they had no other water source except shallow tubewells which did not fall under their preference. For Bihulians, deep groundwater was still a transitional source, because they were shopping around for another surface source. However, they too were not in favour of installing shallow tubewells if they could avoid this.

Some deep tubewell committees introduced wider mechanisms of water control and charging included: adjusting rules for water demand; setting up fees for different activities like use of canal embankments; charging for use of groundwater for fish ponds; charging for transportation of shallow groundwater through deep tubewell canals, mobilising labour resources for deep tubewell cleaning; printing and using vouchers to reduce cheating on paying of dues; and creating rules to control the installation of shallow tubewells.

Conjunctive use between other sources and deep tubewells was possible only when the farmers were able to maintain the deep tubewells. Turnover of tubewells involved setting up of WUGs and training them for various management activities. It was expected that the farmers, with the help of certain guidelines and support for repair and maintenance, would ultimately develop 'sustainable' groundwater irrigation management systems. The farmers did take the help of these guidelines. However, the management of deep tubewells depended on many factors besides those usually outlined in training manuals and brochures. Ironically, the existing knowledge on the management of kulos and jharans, their congruence with the village society and resultant social and political relations played a large role in helping the farmers in certain deep tubewells to work out strategies for deep tubewell management.

Local political units too played a role in the management of irrigation in Tikuligarh and Madhaulia. In Tikuligarh, the VDC played an active role in deep tubewell management and set up a fund to support all the deep tubewells in the VDC. Such type of large-scale support was not evident in the Madhaulia VDC, although they did, try to support one deep tubewell in a crisis. The leaders at the ward level had an important role in deep tubewell and kulo management. Formulating rules and regulations for managing multiple sources and harnessing funds were also easier when the boundaries of deep tubewell, village kulo and the political and administrative units were congruent. The case studies showed the presence of complex sets of laws integrating, for example: traditional rules and regulations of kulos and jharans; deep tubewell rules that were formulated by the project; new rules devised by the farmers and VDCs for managing the tubewells, and emergent rules for conjunctive use. Farmers in the deep tubewell area were not simply guided by the rules set up when they were handed over the deep tubewells. They made their choices from the various 'bundles of rights' pertaining to the various sources.

This study thus showed the dynamic nature of irrigation institutions that are always undergoing transformations. Such transformations take place both within individual irrigation systems and across systems. Such dynamism has to be expected in areas with complex water resources. This dynamism is caused by: the choices created by technological interventions and the opportunities and constraints imposed by them; the changes in flow regime in local rivers; choices of cropping patterns by the farmers, and wider transformations in social relations.

Shifting between groundwater and surface sources

In all cases, farmers have been strategically making choices about using different sources of water at different points in time. Groundwater options first created shifts from use of kulo water. The impact of this was greatest in Mahuwari village, where the migrant population opted for shallow tubewell irrigation for cash-cropping, and have never re-considered a return to use of surface sources. On the other hand, the transformation process underwent different phases in the other villages. Introduction of secure rights to deep tubewell irrigation led the farmers in some villages to cut off ties in surface irrigation (Tikuligarh gaon and Bihuli). The same deep tubewell later became a source of insecurity which they had to manage it themselves. This stimulated some villages to regain water rights from kulos (as in Bihuli), and induced some farmers to install shallow tubewells (as in Tikuligarh gaon).

The transfer process of deep tubewells initially induced farmers in different villages to strengthen their rights to surface sources. The ability of a village to hold on to its surface water rights also depended on its ability to supply labour to the irrigation system. This was evident when one mauja left kulahi in 2003 and opted for groundwater irrigation instead. The process of shifting sources occurred both within individual irrigation deep tubewell systems, and across groundwater and surface irrigation in Tikuligarh, Madhaulia.

I have not examined how such a process affected the management of larger kulo networks. This remains an issue for future research. However, it is clear that it induces adjustments and re-adjustments in management and water allocation. These large farmer-constructed and-managed irrigation systems rely on high input of manual labour for system maintenance and rehabilitation. When portions of irrigation systems leave the network it will make a difference to the way management activities are carried out and also affect the amount of water available at each village outlet. When villages rejoin with the larger networks, the manual labour needed for the maintenance of these systems increases. This study also shows the immense capacity of kulos to incorporate uncertainties. It shows how the kulos have institutionalised a system for facing such changes, in the form of set rules regarding joining, re-joining and leaving the irrigation systems.

Studies conducted in different kulo systems in Nepal have reported dynamism in water rights in the case of kulos sharing a common source of water. Each kulo system employs different strategies to gain control over water. These include shifting intakes, accessing more than one source of water, and making new alliances with other villages while dropping older allies (Gautam, 1994; Shukla 1997). Shukla et. al (1997) reported such dynamism in water rights in studies on several kulo systems in Chitwan District. Such dynamism, according to them can also be related to both external and internal forces. They cite natural forces like floods and changes in stream-flow, different structural forces of intervention processes and internal forces such as increase in population and cultivated area, and changing social relations and power structures as contributing to this dynamism. The findings in this study also concur with these reports.

The findings in this study stress the need to consider how existing surface irrigation could affect groundwater irrigation, and how intervention in groundwater may affect surface irrigation, when designing interventions. This was evident in Madhaulia, where farmers using deep tubewells took the opportunity of later government assistance programme for surface irrigation, and later stopped irrigating from deep tubewells.

Local political dynamics in securing water control

The findings from the case studies show the important role that social and political relations at the local-level play in shaping the emergence of different water use complexes. Everyday relations in society, the role of local level politicians in working out ways for irrigation management and funding, governance at the local level and emergence of other forms of relations related to agricultural production have all played a role in influencing options and choices for different sources of water.

This was very evident in the case of Tikuligarh gaon, where village politics were related with settlement processes there. The power differences between older inhabitants and new settlers played a role in the way farmers opted for different sources of water at different points in time. Intervention processes in groundwater irrigation helped reproduce both competitive as well as cooperative forms of relationships around different sources of water. Intervention in deep tubewell irrigation in an already conflictive atmosphere around the village kulo at first helped a certain group of farmers to strengthen their power over the new water resource. However, the same deep tubewell became a source of conflict after it was transferred back to the farmers. Inegalitarian land holdings and difference in power structures between the irrigators, and locational difference in access to irrigation had been a source of conflict before the deep tubewell was installed here. Farmer behaviour after transfer was again influenced by similar forces amongst the same group of farmers who had taken the support of deep groundwater and slowly ignored participating in surface irrigation. Shallow tubewell technology, was also used by some farmers (who could afford to install it) as a political tool for reducing their share of payment to the deep tubewell and to play politics with those elected as tubewell managers. Similar processes of intervention and same deep tubewell design had different effects in Supauli and Bihuli. In the former, a certain control over water could be worked out by leaders in the village. Deep tubewell technology did provide a constraint to them, but the leaders were able to enforce control over the use of both sources of water.

One of the crucial findings of the study concerns the important roles that certain actors play in shaping the course of water resources management at the local level. Sources such as kulos, jharans and deep tubewells require a body to oversee their management. The role of key actors in their efforts to bring continuity to irrigation after the transfer of deep tubewells is very important. The deep tubewells needed dynamic and even ambitious actors who took over responsibility for managing the deep tubewells after transfer, were involved in working out innovative strategies for managing them, and actively involved in seeking external support strategies for governance over both deep groundwater and surface water.

Implementation of irrigation management transfer (IMT) in Nepal coincided with the period of democratisation in the country. Formal organisations like the water users' groups provided a means for new ambitious politicians to form a base at grass roots level. Management transfer of tubewells was resisted through an organised movement. Water users' group existed as formal bodies with an executive committee in charge of managing the deep tubewells. They were built on the assumption that they are 'apolitical' group of people fully committed to deep tubewell irrigation only.

In practice, handing over deep tubewells was almost like handing it over to local politicians. Keeping in view the demands of a deep tubewell for its operation and maintenance, most of the farmers stated that they did not want to take over their management. The situation has remained the same several years after transfer. It was left to existing or would-be political representatives to take over the management. The committees changed only if a feeling of mistrust by farmers developed towards the executive committee or in tubewells where management transfer started in a conflictive environment right from the start. However, positive changes also started taking place in new tubewells like Durganagar. Here, more farmers were willing to be involved in management after experiencing success in the initial years of management. Another point to note is that the changing political situation in the country and formation of new political parties also drew new people into water users groups to claim it as a platform for political lobbying at local level once again. Overall, such findings suggest that the role of politicians in management of common pool resources is important.

Similar findings were reported by Khanal (2003) in his study on management transfer processes in three surface irrigation systems in the Terai. He shows that water users' associations (WUA) became a platform for political parties to increase their strength in society. Sharma (2001), in his study on rural water supply and sanitation in different areas in Nepal writes that forums like water users' committees (WUC) were becoming the main vehicle for delivering 'development' services in the villages and for strengthening a party's political hold at the local level.

The role of such leaders cannot just be reduced to their role of selfish individuals trying to gain some form of political hold through water users committees. As leaders in the villages, it is their duty to oversee the overall welfare of their jurisdiction including its water. Their involvement was not only in groundwater, but also in surface irrigation. The surface sources and their organisation had existed for many years. The transfer process re-directed the interests of farmers towards these surface sources. Not all leaders held positions for kulo management, but as village leaders they had to ensure proper water delivery into their respective villages.

Sometimes they made use of their different roles to enforce certain decisions. Joint rule making in Supauli proceeded by using the power of an elected village leader rather than simply in terms of the role of kulo chairman or deep tubewell chairman. Case studies also showed that the involvement of village leaders in water management was greater when the boundaries of the mauja, deep tubewell and ward boundaries were congruent. This made it easier for them to build up their political hold in their jurisdiction as well as manage both sources of water.

The leaders developed or made use of different types of social networks to negotiate surface water rights and get external support for deep tubewells. Multiple networks like those based on party politics, kinship relations, surface irrigation and groundwater irrigation were made use of in the process of incorporating different changes.

Agriculture productivity and development

There was a large variation in cropping patterns within and

between the sites. Cash cropping was triggered in Mahuwari because the situation provided for better livelihood prospects for both smallholders and landless villagers. These new opportunities were created by the transforming access to land and markets in this border zone, and flexibilities of shallow tubewell technology. Even though placed within the same market zone, the farmers in the other two sites did not cultivate cash crops in the same intensity as those in Mahuwari. This was related to the socioeconomic position of the farmers. Not all the people in Durganagar did personally cultivate vegetables, even when the operating routine of the deep tubewell was conducive to vegetable farming. Those people who were involved in small businesses and services and did not have enough time for farming, gave out a portion of land for sharecropping. This was mostly for vegetable cultivation. Tikuligarh farmers, who were located within the command areas of older tubewells, irrigated vegetables by means of shallow tubewells also.

After incorporating groundwater as an additional input, the farmers had developed different methods of sharing costs incurred in agricultural production. Two types of relations in agricultural production existed between landowners and cultivators: sharecropping and the contract system. While sharecropping was the common practice in Tikuligarh and Madhaulia, farmers in Mahuwari were involved in both sharecropping and contract systems. The contract system was mainly followed in banana cultivation.

Cost sharing under sharecropping entailed a fifty-fifty share of the produce between the landowner and the cultivator. This was the old practice when only kulos were used for irrigation. The same practice was carried over when groundwater was added in the farming system. The cost for the flat rate of electricity of deep tubewells was borne by the landowner while the cost of operation shifted to cultivators. This was an added burden to sharecroppers but they continued with groundwater irrigation. Vegetables were the main crop on land given out for sharecropping. The farmers in Mahuwari started practicing contract farming, which was made possible by absentee landlords, availability of shallow tubewells and opportunities for cash cropping. Possibilities of profitable agriculture drew them towards shallow groundwater, which gave them more control over water for irrigation.

Beyond intervention and system models: implications for future water and development policies

Deep tubewell irrigation in Madhaulia and Tikuligarh was mostly driven by programme models for creating infrastructure and 'crafting' specific institutions around the tubewell technologies. They were constructed as public tubewells and rights handed over to the farmers. Dissemination of shallow tubewell technology was, and still is, seen as a medium through which farmers gain control over technology and water resource, but with less specificity on organising institutions. Property rights to shallow groundwater have been created by propagating two types of models: community tubewells and individually owned tubewells. There are different models of group shallow tubewell development in use. One is where tubewells installed through a cluster approach (which is the approach adopted by the government groundwater project in some districts including Rupandehi). Another is through different livelihood programmes like the group tubewell approach in Mahuwari, implemented through community organisations. The other recent model includes that propagated by the 'community groundwater irrigation sector project' in eastern Terai districts that started in 1999. This model takes the VDC as a subproject. It implements group shallow tubewell schemes by creating several water users' groups. In addition to the individual water users' groups, a water users' association is also created. This assists the different water users' groups in different activities related to agriculture (CGISP 2005).

Models imply that 'things can/must be done in a particular way' (Mollinga and Bolding 2004: 293). The notion of replication comes in with the word 'model'. The same model is applied across the area with the assumption of homogeneity in the context that it is supposed to operate in. This approach is found in the case of both deep tubewell and shallow tubewell development. The lessons learnt from a certain model in a certain place should then apply somewhere else. When one examines the history and modes of groundwater development in the study area, it can be said that groundwater irrigation development has already come a long way given the different models of intervention practiced for it. There have been modifications in the design of the different models through the years. However, water resources development of groundwater for irrigation has remained largely a technology-driven endeavour at government level, carried out on the basis of a number of assumptions on manageability and control.

The BLGWIP has often been cited as "one of the successful projects in the Terai after system development and design using farmers' involvement" and that "it had established an effective model and capacity for groundwater development using improved deep tubewell technology and users' participation" (Myint, in Salman 1997: 127). The findings here show that there has been failure to reassess technological relevance (performance) in relation to the need of the farmers.

These case studies show that farmers in study area are involved in creating or maintaining legitimate access and organisation around different combinations of irrigation water sources that they find optimal. In order to understand the reality of groundwater use and management in an area of complex water resources, it is necessary to look beyond the concept of infrastructure system models and isolated technological interventions. Farmer behaviour is very strategic and they choose, accept, adjust or even reject systems when working out the best ways for irrigation that are most effective for local production options. It is clear from the study, that in an area of complex water resources different actors strategise, manoeuvre, and create social relations as they incorporate the heterogeneous nature of interactions between intervention processes in groundwater irrigation and existing systems of irrigation. The area is basically a 'development arena' where different technological innovations are worked out, different strategies formulated and networks created. Linkages between the various sources of water, technologies and people are created in the process. For example, the farmers in Tikuligarh and Madhaulia tried to incorporate the deep tubewells within the existing kulo and jharan systems. In Mahuwari, farmers tried to work out innovative ways to access shallow groundwater.

Negotiations and interactions took place within the domain of villages, wards, VDCs and surface irrigation network. The interactions took place at the level of a deep tubewell as a unit, several deep tubewells together, between deep tubewells and surface sources, between deep tubewells and shallow tubewells, and between several shallow tubewells. In Mahuwari, the farmers were involved in relations and networks that extended from the village to the drilling enterprise and even across the border to India to get the shallow tubewell technology.

Farmers make use of the different technical, organisational, normative/legal options in the process of negotiating their water rights. The complexes of water use are shaped by the performances of technology, social networks and opportunities provided by different sources of water. It also depends on which source of water provides them with better opportunity at that point in time. This strategic behaviour, and the way farmers define property rights is embedded in larger structures. Understanding groundwater use in the study sites was to understand the shaping, becoming and transformation of groundwater technology within the water cycle.

The findings show that some deep tubewells definitely 'perform' better than others. However, the notion of 'model' reduces groundwater use and management to a very de-contextualised form, because even deep tubewells of the same 'model' that lie side by side 'perform' quite differently from each other. For example, while Durganagar tubewell is one of the tubewells in the whole project area with the highest use hours, a tubewell of the same design or of the same model is not used much by the farmers. Technical changes in design and approaches to implementation did lessen many problems for the farmers in terms of management, but the study shows that tubewells of any kind - deep or shallow - do not function as designed. Rather, they become part of the whole existing context in the area in which they are inserted into.

The handover process for deep tubewells has meanwhile, followed a policy prescription on irrigation management transfer. A number of deep tubewells were turned over to the farmer groups in a similar manner. Each tubewell unit of the same generation was considered similar in the process. It has been one of the crucial factors shaping the course of irrigation development in the area. This created a multitude of institutions that started to 'piggy-back' on other institutions for support. The 'second generation' issue or problem of irrigation management transfer was the creation of deep tubewell irrigation institutions that leaned heavily on political structures for sustenance. Tikuligarh VDC was very responsive to creating support mechanisms by creating a fund for all deep tubewells within its jurisdiction. The relative high income of this VDC and the strategies of the leaders helped to build this up. In Madhaulia, investing in a dam was the preferred choice. This was possible because of the presence of an alternative (surface) source of water. Rather than moving towards the making of a viable 'sustainable' deep tubewell water users' group, irrigation management reverted back to issues of security and insecurity in surface irrigation. The strategies of leaders in certain deep tubewell areas to re-organise a larger network of deep tubewells in an effort to form a platform from where to rally 'for support is also an outcome of the feeling of insecurity created by the turnover process. The sustenance of deep tubewells, supposed to have been realised through returns from agriculture, was in fact depending on the skills and capacities of local leaders to find investments from different means for operation of the tubewells.

The term conjunctive use, as has been mentioned in chapter one is a term that has been used quite frequently in irrigation policy making. It had gained more significance after the implementation of Agriculture Perspective Plan with its focus on expanding groundwater services in the Terai. Irrigation Policy (2003) specifically mentions the 'need to promote conjunctive use of groundwater and surface water irrigation system' in its rationale. The policy, in its section on 'study, identification and selection of the projects' states that the basis for selection of projects would be based on the 'feasibility for conjunctive use of surface and groundwater', and that the 'concerned stakeholders' would be 'coordinated in the process of project selection' (Irrigation Policy 2003: 4). This study shows the challenge this involves. Selecting projects based on feasibility of conjunctive use could thus mean several things. For instance, it could mean that one could select projects (groundwater?) based on the fact that there is already either a seasonal or permanent shortage in supply of surface source (which is the most likely situation). Or it could mean designing a whole new project that incorporates the use of both the sources of water. It could even mean, constructing a surface irrigation system in order to make up for constraints in flexible delivery found in old deep tubewell irrigation systems (as the cases in this study show). The findings from the case studies throw insights into strategies worked out by farmers for such systematic conjunctive management.

The findings of this study clearly show that there is a need to look beyond interventions, and to focus on local action processes, in order to understand both resource use and technological performance, and future options in their transformation. Such processes, not usually discussed in policy making and designs of projects, can have power to modify water use practices of other irrigation systems in the vicinity. Farmers' choices for crops depend on both factors of input and market and also on the availability of labour. Low levels of use are not just related to difficulties to form groups but are also related to costs and multiple choices. This brings in variation in the performance of 'irrigation systems'. Problems and issues in groundwater irrigation, as well as prospects in conjunctive use, cannot be understood only based on one water source only.

The study proves the advantage of understanding complexes of water use and the social spaces that they function in. Without such understanding, it becomes difficult to answer the real issues that irrigation management entails and which are often just listed as problems. For example, the use of deep tubewells did decrease considerably after transfer. From a policy point of view, this was a 'problem' of 'under-utilisation' or 'low performance' of groundwater. For the farmers, the transfer policy was expensive. Hence they tried to mitigate this by optimising the use of groundwater and integrating all sources as effectively as possible.

The linkages between the uses of various sources of water had to be understood in order to understand the reality of irrigation practices in these groundwater intervention areas. In order to do so, the different isolated technological interventions in irrigation and existing forms of water use systems were placed within the broader water resources systems they are part of and function in. They were understood as 'intricate complexes of physical-technical, organisational and normative-legal dimensions of water control' within a water resource system and that they developed in the larger agro-ecological, social, political, cultural and technological context. Through this approach it is possible to understand groundwater use, when it is used alone and also in conjunction with other sources of water. Conjunctive management could thus be understood as an intricate complex of physical-technical, organisational and normative-legal dimensions of water control that comes up through interactions between various sociotechnical complexes (each with their own physical-technical, organisational and normative-legal dimensions) in a water resource system. It develops in a wider politico-economic, socio-economic and socio-cultural context in which all these complexes function together.

The linkages between the use and management of various sources of water is understood by looking at the technical, normative and organisational interdependence within and between them and how they are shaped by the larger hydrology and social processes of the area. Diversity of rule making was examined from the perspective of legal pluralism. Intervention processes in irrigation interact with, and are embedded in the already existing power relations at the local level. They 'interact with already contested domains and power and meaning' (Li 1996: 515 in Mosse 1997: 499). Intervention in deep tubewell irrigation helped change social relations between different people in the villages. Shallow tubewells meanwhile added on to the process of bringing forth either more co-operative or conflictive relationships. By analysing these processes as embedded in larger structures, farmers' actions and the way they define property rights can be better understood.

I preferred not to conceptualise water use in a complex river basin with multiple interventions in terms of any fixed boundaries defined through technological interventions like deep tubewells, or even those 'formally' defined by the farmers, as with the FMISs. When farmers use water from different sources there is a blurring of these boundaries. The water use complexes showed how farmers inside a VDC were grouped together through common sources of water. This complex formed the dynamic hydrological boundary that they had defined. Yet, taking the village or VDC as the focus of study made it possible to locate these water complexes within both physical and social space, as well as part of the larger water resource system. This study approach shows that hydrological and water management studies should give more attention to social space and not just the hydrological or technological system only.

This book tried to capture how interventions in groundwater irrigation interplayed with existing management practices of other sources of water and how the management practices of groundwater itself have been influenced in the process. It did so in case of interventions in both deep tubewell as well as shallow tubewells. In recent years, development approaches have focused on the process of decentralization and empowerment of local political bodies like the District Development Committees (DDC) and the VDCs. These have been given more power in decision making in development activities and therefore also on the control of natural resources within their jurisdiction. The case studies in Tikuligarh and Madhaulia showed that the role of the local-level governance structures in decision making for irrigation management and governance is prominent and needs more recognition. Their role was significant in the case of deep tubewell and surface irrigation management. The case of Mahuwari dealt with farmers own initiatives in shallow tubewell installations with minimum role from the VDC. However, different types of intervention programmes both in groundwater (deep and shallow) as well as for other sources of water for irrigation implemented through the VDC have to have to take into account the potential role of the local level governance structures in decision making and control of the natural resources. A total of 181 deep tubewells were installed throughout the district. A further study in other VDCs inside the district itself or in other districts in the Terai, with and without alternate sources of water would shed light on this matter. It is also important to understand the same in case of shallow tubewell irrigation development. This study was not able to capture this dynamic in more detail because most of the tubewells installed in the study sites had been installed through the previous programmes by the ADBN and privately at their own cost by the farmers. Recent programmes in shallow tubewell irrigation, especially those involved in installing group shallow tubewells are being implemented through the VDC. Further studies could also focus on understanding interactions beyond this level in order to understand the interactions within and between different water use systems across the river basin.

References

Abhayaratna, M.D.C., Douglas Vermillion and Johnson S. Perry. eds. 1994. *Farmer management of groundwater irrigation in Asia.* Selected papers from a South Asian regional workshop. Colombo, Sri Lanka. IIMI.

Adhikari, Jagannath. 1996. *Biased agrarian restructuring: the beginnings of agrarian change, a case study in central Nepal.* T.M. Publications.

Adhikari, Madhukar and Rajendra Pradhan. 2000. Water rights, law and authority: changing water rights in the Bhamke khola basin. In *Water, land and law: changing rights to land and water in Nepal,* eds. Rajendra Pradhan, Franz von Benda-Beckmann and Keebet von Benda-Beckmann: 71-99. Legal Research and Development Forum, Nepal, Wageningen University and Erasmus University, The Netherlands. Jagadamba Press, Nepal

Alberts, J. H. 1998. Public tubewell irrigation in Uttar Pradesh, India. A case study of the Indo-Dutch tubewell project. ILRI special report, March 1998. ILRI. Wageningen, Netherlands.

APROSC/JMA. 1995. Nepal agriculture perspective plan. Agricultural Projects Services Centre and John Mellor Associates, Inc.

Ballabh, Vishwa, Kameshwar Choudhary, Sushil Pandey and Sudhakar Mishra. 2003. Groundwater irrigation and agrarian change in eastern India. IWMI Tata 9. Comments by Aditi Mukherji. From http://www.iwmi.org/iwmi-tata.

Basnet, K. 1996. Conjunctive use of ground and surface water for irrigation: case study of an irrigation system in the Terai of Nepal. Unpublished report submitted to Winrock International policy analysis in agriculture and related resource programme, Kathmandu, Nepal.

Benda-Beckmann, Franz, Keebet von Benda-Beckmann and Joep

Spiertz. 1996. Water rights and policy. In *The role of law in natural resources management*, eds. Joep Spiertz, and Melanie G. Wiber: 77-99. 's Gravenhage: VUGA.

Benda-Beckmann, Franz and Keebet von Bend-Beckmann. 1999. A functional analysis of property rights, with special reference to Indonesia. In property rights and economic development. *Land and natural resources in south-east Asia and Oceania,* T.van Meijl and F.von Benda-Beckmann (eds), London: Kegan Paul.

Benda-Beckmann, Franz von. 1997. Citizens, strangers and indigenous peoples: conceptual politics and legal pluralism. *Law and Anthropology* 9: 1-42.

Benda-Beckmann, Franz von. 2002. Who is afraid of legal pluralism? *Journal of legal pluralism* 47:1-46.

Benjamin, Paul. 1994. Historical basis of irrigation in Nepal. In *Institutions, incentives and irrigation in Nepal*, eds. Paul Benjamin, Wai Fung Lam, Elinor Ostrom and Ganesh Shivakoti : 15-67. workshop in political theory and policy analysis, Indiana and Rampur.

Benton, Ted. 1992. Ecology, socialism and the mastery of nature: a reply to Reiner Grundman. In *New Left Review* 1: (194):55-74.

Béteille, Andre. 2002. *Equality and universality: Essays in social and political theory.* Oxford University Press, New Delhi.

Bhairahawa Lumbini groundwater project IDA Credit 654. Project completion report. 1984. World Bank, South Asia Projects Department.

Bhairahawa-Lumbini groundwater irrigation project. 1999a. Stage III. Project completion report April 1999 draft report. Tahal Consulting Engineers Ltd.

Bhairahawa-Lumbini groundwater irrigation project. 1999b. Stage III. Project completion report July 1999 final report. Tahal Consulting Engineers Ltd.

Bhandari, Hum Nath. 1999. Economics of groundwater irrigation in rice-based systems in Terai of Nepal. PhD dissertation, University of the Philippines Los Banos, Philippines.

Bhandari, Ravi. 1999. Land tenure in Nepal: A constraint or catalyst for economic growth? Nepal Tarai rural development options study. World Bank.

Bhandari, Ravi. 2001. Social distance in sharecropping efficiency: the case of two rice-growing villages in Nepal. Phd. Dissertation. University of California at Berkeley, USA.

Bhatia, B. 1992. *Lush fields and parched throats: political economy of groundwater in Gujarat*. Working Paper. World Institute for Development Economics Research, Helsinki, Finland.

Boelens, Rutgerd, Dik Roth and Margreet Zwarteveen. 2002. Legal complexity and irrigation water control: analysis, recognition and beyond. Paper for the 13th international congress on folk law and legal pluralism, Thailand.

Boelens, Rutgerd. 1998. Collective management and social construction of peasant irrigation systems: a conceptual introduction. In *Searching for equity in peasant irrigation*, eds. Rutgerd Boelens and Davila Gloria: 81-99. Van Gorcum, The Netherlands.

Boyce, J. 1987. *Agrarian impasse in Bengal: agricultural growth in Bangladesh and West Bengal*, 1949-1980. Oxford University Press. New York.

Bruns, Randolph B., Meinzen-Dick, R. eds. 2000. *Negotiating water rights*. International Food Policy Research Institute. Washington DC, U.S.A.

Burawoy, Michael. 1985. *The politics of production. Factory regimes under capitalism and socialism*. Verso, London.

Burawoy, Michael and Erik Olin Wright. 2000. Sociological Marxism. Sourced at: www.ssc.wisc.edu/~wright/SocMarx.pdf

Chakravarti, Anand. 2001. *Social power and everyday class relations*. Agrarian transformation in North Bihar. Sage Publications. New Delhi.

Chambers, Robert. 1988. *Managing canal irrigation. Practical analysis from South Asia*. New Delhi and Calcutta: Oxford & IBH Publishing Corporation.

Chaudhary, Shankar Lal. 2003. *Tharus: The pioneer of civilization of Nepal*. Muna Press. Lalitpur, Nepal.

Clay, E.J. 1974. Innovation, inequality and rural planning: the economics of tubewell irrigation in the Kosi region, Bihar, India. Unpublished Ph.D. Thesis. University of Sussex.

Delft Hydraulics. 1988. *Water intake design review, draft final review*, United Nations Development Programme office for project execution.

Department of Hydrology and Meteorology. 1971-2001. *Climatological records of Nepal*.

Desai, M., Susanne Hoeber Rudolph and Rudra, A. eds. 1984. *Agrarian power and agricultural productivity in South Asia*. Oxford University Press. New Delhi.

Development Program 1999. Ek jhalak. District agriculture development office, Rupnadehi District.

Dhawan, B.D. 1982. *Development of tubewell irrigation in India.* Institute of Economic Growth, Delhi. Agricole Publishing Academy.

Dixit, Kanak Mani and Shastri Ramachandran eds. 2002. *State of Nepal.* Himal Books. Kathmandu.

Dowding, Keith. 1996. *Power.* Open University Press, UK.

Dubash, Navroz. K. 2002. *Tubewell capitalism: groundwater development and agrarian change in Gujarat.* Oxford University Press, New Delhi.

Gautam, Suman Rimal and Prem B. Shrestha. 1997. Technological constraints to the optimum utilization and expansion of groundwater irrigation in Nepal Tarai. Research report series no. 38, Winrock International, December, Kathmandu.

Gautam, Suman Rimal. 1994. Irrigation management in a mixed community: a case study in Chitwan, Nepal. M.E. Thesis. School of Civil Engineering, Asian Institute of Engineering, Bangkok, Thailand.

Giddens, Anthony. 1984. *The constitution of society: outline of the theory of structuration.* Berkeley and Los Angeles. University of California Press.

Gill, Gerard J. 2003. Seasonal labour migration in rural Nepal: A preliminary overview. Overseas Development Institute. Working Paper 218. London.

Gorter, Pieter. 1989. Canal irrigation and agrarian transformation: a case of Kesala. *Economic and Political Weekly*, 24(39): A94-A105.

Groundwater Development Consultants Ltd. Cambridge, United Kingdom and Hunting Technical Services Ltd and East Consult (P) Ltd. 1994. Reassessment of the groundwater development strategy for irrigation in the Terai. Volume 1. Main report. HMG Nepal, department of irrigation, groundwater resources development project. Kathmandu Nepal

Guillet, David. 1992. *Covering Ground: communal water management and the state in the Peruvian highlands.* University of Michigan Press. Ann Arbor, Michigan. United States.

Gyawali, Deepak and Ajaya Dixit. 1999. Fractured institutions and physical interdependence: challenges to local water management in the Tinau river basin, Nepal. In *Rethinking the mosaic. investigation into local water management*, eds. M. Moench, E. Caspari and Ajaya Dixit: 55-121. Kathmandu and Colorado. Nepal Water Conservation Foundation, Institute for Social and Environmental Transition.

His Majesty's Government. 2001. Water resources strategy formulation. Water and Energy Commission Secretariat. Kathmandu.

Hoogesteger van Dijk, J. 2005. Making do with what we have: understanding drought management strategies and their effects in the Zayandeh Rud Basin, Iran. M.Sc. Thesis. Irrigation and Water Engineering Group, (IWMI),Wageningen University, the Netherlands.

IDRC. 2005. Definition of conjunctive use from http://www.idrc.ca/en/ev-29788-201-1-DO TOPIC.html.

IIMI, 1991. Process and performance evaluation of ADBN supported irrigation schemes. Volume I. Main Report.

Irrigation Policy. 2003 (2060 B.S.). From the website of HMG Nepal, Ministry of Water Resources, Department of Irrigation, Kathmandu.
http://www.doi.gov.np/acts/irrigation_policy.pdf.

Japan International Cooperation Agency. 1992. The master plan study on the Terai groundwater resources evaluation and development project for irrigation, Interim Report. Kathmandu, Nepal.

Jorgensen, U. and Ole Henning Sorensen. 1999. Arenas of development: a space populated by actor-worlds, artefacts and surprises. In *Technology Analysis and Strategic Management.* 11 (3).

Kahnert, F. and G. Levine, eds., 1993. Groundwater irrigation and the rural poor: options for development in the Gangetic Basin. A World Bank Symposium.

Kansakar, Dibya R. 1997. Groundwater resources. Paper presented at the Nepal-water resources strategy formulation workshop. HMG Nepal World Bank /JCF.

Khanal Puspa R. 1994. Integrated use of surface and ground water for irrigation in Narayani zone, Nepal. M.E. Thesis. Asian Institute of Technology, Thailand.

Khanal, Puspa R. 2003. *Engineering Participation: the processes and outcomes of irrigation management transfer in the Terai of Nepal.* Wageningen University Water Resources Series 2. Hyderabad. Orient Longman.

Kishore, Avinash. 2004. Understanding agrarian impasse in Bihar. *Economic and Political Weekly*, 39(31):3484-3491.

Kloezen, W. and Peter P. Mollinga. 1992. Opening closed gates: recognizing the social nature of irrigation artefacts. In: G. Diemer and J. Slabbers, eds. *Irrigators and engineers.* Amsterdam:

Thesis Publishers: 53-64.

Kloezen, W. H. 2002. Accounting for water: Institutional viability and impacts of market-oriented irrigation interventions in central Mexico. PhD thesis. Wageningen University, The Netherlands.

Knegt. Jan-Willem F. and Linden F. Vincent. 2001. From open access to access by all: restating challenges in designing groundwater management in Andhra Pradesh, India. *Natural Resources Forum.* 25: 321-331.

Knegt. Jan-Willem F. 2000. Murky waters. Control over groundwater extraction for small scale irrigation in a village in southern Andhra Pradesh, India: the role of tubewell technology and electricity supply. M.Sc. Thesis. Wageningen University, The Netherlands.

Koirala, Govind P. 1998. Clogs in shallow groundwater use. Research report series no. 39. Winrock International, Kathmandu, Nepal.

Koirala, Govind P. 2001. http://www.cgisp.org.np/sub.html.

Koirala, Govind P. and Ganesh Thapa. 1997. Food security challenge: Where does Nepal stand? Research report series no. 36. Winrock International. Kathmandu, Nepal.

Koirala, Govind P. and Suman Rimal Gautam. 1998. Shallow tubewell irrigation in Nepal: expansion and utilisation issues: Policy outlook series no. 3. Winrock International. Kathmandu, Nepal.

Lam, Wai Fung. 1998. *Governing irrigation systems in Nepal: institutions, infrastructure and collective action.* Oakland, California: Institute for Contemporary Studies.

Leeuwen van Mathij. 1998. Fluid institutions. M.Sc. Thesis. Wageningen University, The Netherlands.

Long, N, J.D. van der Ploeg, Chris Curtin and Box, Louk. 1986. The commoditization debate: labour process, strategy and social network. *Papers of the Department of Sociology 17.* Agricultural University Wageningen, The Netherlands.

Long, Norman. 2001. *Development sociology: actor perspectives.* London. Routledge.

Long, Norman and J.D. van der Ploeg. 1989. Demythologizing planned intervention: an actor perspective. *Sociologia Ruralis.* 29(3- 4): 226-249.

Long, Norman ed. 1989. *Encounters at the interface. A perspective on social discontinuities in rural development.* Wageningen Studies in

Sociology 27. Wageningen Agricultural University, The Netherlands.

Magadlela, Dumisani. 2000. Irrigating lives: development intervention and dynamics of social relations in an irrigation project. Phd thesis, Wageningen University, The Netherlands.

Manzungu, Emmanuel. 1999. Strategies of smallholder irrigation in management in Zimbabwe. PhD thesis, Wageningen University, The Netherlands.

Martin, Edward D. 1986. Resource mobilization, water allocation, and farmer organization in hill irrigation systems in Nepal. Ph.d dissertation. Cornell University.

McCay, J. B. 2002. Emergence of institutions for the commons: contexts, situations, and events. In *The drama of the commons*, eds. Ostrom, E., Thomas Dietz, Nives Dolsak, Paul C. Stern, Susan Stonich and Elke U. Weber : 361-402. National Academy Press. Washington DC.

Mehta, Lyla. 1998. Context of scarcity: the political ecology of water scarcity in Kutch, India. Phd dissertation. Sussex, UK. Institute of Development Studies.

Meinzen-Dick, R. 1996. Groundwater markets in Pakistan: Production and productivity. Research report 105. International Food Policy Research Institute. Washington DC, U.S.A.

Meinzen-Dick, Ruth S. and Rajendra Pradhan. 2001. Implications of legal pluralism for natural resource management. *IDS Bulletin*. Vol 32(4):10-17.

Ministry of water resources. 1990. *Irrigation Master Plan*. HMG of Nepal, MOWR, Kathmandu.

Ministry of water resources. 1997. *Irrigation Policy 2049 (1992). First Amendment, 2053 (1997)*. HMG of Nepal, MOWR. Kathmandu.

Ministry of water resources. 1997. *Water Resources Act 2049 (1992)*. HMG of Nepal, MOWR, Kathmandu.

Ministry of water resources. 1997. *Water Resources Regulation 2050 (1993)*. HMG of Nepal, MOWR, Kathmandu.

Moench, Marcus. 1994. Selling water: conceptual and policy debates over groundwater markets in India: papers of the workshop on water management: India's groundwater challenge held at VIKSAT. Ahmedabad.

Moench, M.H. and J.J. Burke. 2000. *Groundwater and society: resources, tensions and opportunities: themes in groundwater management for the twenty-first century*. United Nations department of economic and social affairs. United Nations Publication.

Molle, François. Defining water rights: by prescription or negotiation? *Water policy* 6: 207-227.

Mollinga, Peter P. 1998. On the waterfront. Water distribution, technology and agrarian change in a South Indian canal irrigation system. PhD thesis. Wageningen University, The Netherlands.

Mollinga, Peter P. and Alex Bolding, eds. 2004. *The politics of irrigation reform: contested policy formulation and implementation in Asia, Africa and Latin America. England.* Ashgate Publishing Limited.

Mosse, David. 1997. The symbolic making of a common property resource: history, ecology and locality in a tank-irrigated landscape in south India. *Development and change.* 28: 467-504.

Mosse, David. 2004. Is good policy unimplementable? Reflections on the ethnography of aid policy and practice. *Development and change.* 35: 639-671.

Myint, Ohn. 1999. Experience with groundwater irrigation in Nepal. In Groundwater: legal and policy perspectives: Proceedings of World Bank Seminar eds. Salman M.A. Salman: 125-132. The World Bank, Washington D.C., U.S.A.

O'Mara, Gerald T. 1988. The efficient use of surface water and groundwater in irrigation: an overview of the issues. In efficiency in irrigation: the conjunctive use of surface and groundwater resources, eds. O'Mara, Gerald T: 1-16. A World Bank symposium. The World Bank, Washington D.C., U.S.A.

Oad, R., and R.K. Sampath 1995. Performance measure for improving irrigation management. *Irrigation and Drainage Systems*, 9: 357-370.

Olin, M. 1994. Transfer of management to water users in stages I&II of the Bhairahawa Lumbini groundwater irrigation project in Nepal. Paper presented at the International Conference on Irrigation Management Transfer: 20-24 September 1994. IIMI and Wuhan University of Hydraulic and Electrical Engineering, China.

Onta, P.R., Das Gupta, A. and Poudyal, G.N. 1991. Integrated irrigation development planning by multi objective optimization, *Journal of Water Resources Planning and Management* 117 (6): 662-678.

Oorthuizen, J. 2003. *Water, works and wages: the everyday politics of irrigation management reform in the Philippines.* Wageningen University Water Resources Series 3. Hyderabad. Orient Longman.

Ostrom, Elinor. 1990. *Governing the commons: The evolution of institutions for collective action.* New York: Cambridge University Press.

Ostrom, Elinor. 1992. *Crafting institutions for self-governing irrigation systems.* San Francisco: Institute for Contemporary Studies Press.

Palmer-Jones, R.W. 1994. The turnover of deep tubewells for irrigation. International conference on irrigation management transfer. Draft Conference Papers. IIMI and Wuhan University of Hydraulic and Electrical Engineering, China.

Pandey, Bikash. 2003. Nepali Times edition 23 October. People Power.

Pant, Dhruba Raj. 2000. Intervention processes and irrigation institutions: sustainability of farmer managed irrigation systems in Nepal. Phd thesis. Wageningen University, The Netherlands.

Pant, Niranjan and R.P. Rai. 1985. *Community tubewell and agricultural development: an organizational alternative to very small farmers irrigation in east Gangetic Plains.* Ashish Publishing House. New Delhi, India.

Pant, Niranjan. 2004. Trends in groundwater irrigation in eastern and western UP. *Economics and Political Weekly* 39(31)3463-3468.

Parajuli, Umesh N, Chiranjivi Sharma, Suman Rimal Gautam and Ram Prakash Yadav. 2001. Hill irrigation: the question of rehabilitation. Research report Series No. 46. Winrock International. Kathmandu.

Parajuli, Umesh N. 1999. Agro-ecology and irrigation technology: comparative research on farmer-managed irrigation systems in the mid-hills of Nepal. Phd thesis. Wageningen University, The Netherlands.

Paudyal, G.N. and A.D. Gupta 1987. Operation of a groundwater reservoir in conjunction with surface water. *International Journal of Water Resources Development.* 3 (1): 31-43.

PDDP, 2000. Baseline survey of Hatti Bangain village development Committee: surveyor: Ms. Sarita Koirala social mobiliser for the VDC. PDDP/UNDP, unpublished report. Rupandehi, Nepal.

Poudel, Som Nath. 2000 Investment in irrigation in Nepal: outlook for the future, draft report. Winrock International. Kathmandu

Prachanda Pradhan. 1989. *Patterns of irrigation organization in Nepal: a comparative study of 21 farmer-managed irrigation systems.* IIMI, Colombo, Sri Lanka.

Pradhan, Rajendra, Franz von Benda-Beckman, Keebet von Benda Beckmann, H.L.J.Spiertz, Shantam S. Khadkha and K. Azharul

Haq. 1996. *Water rights, conflicts and policy.* Proceedings of a workshop held in Kathmandu, Nepal, January 22-24, 1996. International Irrigation Management Institute.

Pradhan, Rajendra and Ujjwal Pradhan. 1996. Staking a claim: law, politics and water rights in farmer managed irrigation systems in Nepal. In *The role of law in natural resources management,* eds. Joep Spiertz and Melanie G. Wiber. VUGA: 61-76. The Netherlands.

Pradhan, Rajendra. 2000. Land and water rights in Nepal (1854-1992). In *Water, land and law: changing rights to land and water in Nepal,* eds. Rajendra Pradhan, Franz von Benda-Beckmann and Keebet von Benda-Beckmann: 39-70. Proceedings of a workshop held in Kathmandu, 18-22 March 1998 Freedeal, Wageningen Agricultural University and Erasmus Universiteit Rotterdam.

Pradhan, Ujjwal. 1990. Property rights and state intervention in hill irrigation systems in Nepal. Ph.D. dissertation, Cornell University.

Prakash, Anjal. 2005. *The dark zone: groundwater irrigation, politics and social power in North Gujarat.* Wageningen University Water Resources Series 7. Hyderabad. Orient Longman.

Rana, Purushottam Shumshere Janga Bahadur Rana. 2003. *Shree teen haruko tathya brithanta, bhag-2.* Bidyarthi Pustak Bhandar, Kathmandu.

Regmi, Mahesh C. 1971. *A study in Nepali economic history,* 1768-1846. New Delhi: Manjusri.

Regmi, Mahesh C. 1978. *Thatched huts and stucco palaces: peasants and landlords in 19th Century Nepal.* New Delhi: Vikas.

Regmi, Mahesh C. 1976. *Landownership in Nepal.* Berkeley: University of California Press

Rose, Leo E and John T. Scholz. 1980. *Nepal: Profile of a Himalayan Kingdom.* Boulder, Colorado, Westview Press.

Roth, Dik. 2003a. Ambition, regulation and reality. Complex use of land and water resources in Luwu, South Sulawesi, Indonesia. Ph.D thesis. Wageningen, The Netherlands: Wageningen Agricultural University.

Roth, Dik. 2003b. Which order? Whose order? Local irrigation management in a Balanise migrant society in Sulawesi, Indonesia. Paper presented at the workshop 'Order and Disjuncture: the organisation of aid and development,' SOAS, London, 26-28 September 2003.

Saxena, R.S. and Prakash Singh.1988. Groundwater management in eastern Uttar Pradesh: A study of underdevelopment. Paper presented at the workshop on development and management of groundwater resources in eastern Uttar Pradesh, 5-6 April, Narendra Dev University of Agriculture and Technology, Faizabad, India.

Schlager, E. and Elinor Ostrom. 1992. Property-rights regimes and natural resources: a conceptual analysis. *Land Economics* 68(3):249-262.

Shah, S.G. and Gautam Narayan Singh. Irrigation development in Nepal, investment, efficiency and institution. December 2001. *Research report series No. 47. Winrock International*, Kathmandu.

Shah, Tushaar. 1993. *Groundwater markets and irrigation development. political economy and practical policy*. Oxford University Press.

Shaha, Rishikesh. 1996. *Modern Nepal. A Political History 1769-1955.* Volume 1 and 2. Manohar Publishers, India.

Shamima Siddika.1993. *The Muslims of Nepal.* Gazala Siddika.

Sharma, C.K. 1978. *Natural Resources of Nepal.* Navana Printing Works, Calcutta, India.

Sharma, C.K. 1981. *Groundwater Resources of Nepal.* Navana Printing Works, Calcutta, India.

Sharma, Sudhindra. 2001. *Procuring Water: foreign aid and rural water supply in Nepal.* Nepal Water Conservation Foundation. Format Printing Press, Kathmandu, Nepal.

Shivakoti, Ganesh. 1992. Variation in interventions, variations in results: assistance to farmer-managed irrigation systems in Nepal. Irrigation Management Network, Paper 11. Overseas Development Institute, London, England.

Shrestha, Ratna S. and Nirmal Kumar Sharma. 1987. A comparative study of farmer-managed and agency-managed irrigation systems. In Irrigation management in Nepal. Research papers from a national seminar. International Irrigation Management Institute; Institute of Agriculture and Animal Sciences and Winrock, Kathmandu, Nepal.

Shrevel, A. eds. Groundwater management: sharing responsibility for an open access resource. Proceedings of the first Wageningen Water Workshop. 13-15 October 1997. *ILRI Special Report*, December 1997. ILRI, Wageningen, The Netherlands.

Shukla, Ashutosh, Ganesh Shivakoti, Rabi Poudel and Niraj N. Joshi. 1996. A process documentation study explaining dynamism in water rights in East Chitwan. *In People and participation in sustainable development: understanding the dynamics of natural resource systems*, eds. Ganesh Shivakoti, George Varughese, Elinor Ostrom, Ashutosh Shukla and Ganesh Thapa: 146-159. Proceedings of an international conference held at the Institute of Agriculture and Animal Science, Rampur, Chitwan, Nepal, 17-21 March, 1996. Bloomington, Indiana and Rampur, Chitwan, Nepal.

Siddiq, A. 2001. From http://www.cgisp.org.np/sub.html.

Skerry C.A., Kerry Moran, Kay M. Calavan. 1991. *Four decades of development: the history of U.S. assistance to Nepal (1951-1991)*. USAID. Kathmandu Nepal.

Spiertz, H.L.J. 2000. Water rights and legal pluralism: some basics of a legal anthropological approach. *In negotiating water rights*, eds. B.R. Bruns and R. S. Meinzen-Dick: 245-268. IFPRI. New Delhi.

Steenbergen, Frank van. 1997. Institutional change in local water resource management: cases from Balochistan. *Netherlands Geographical Studies 220*. Utrecht University, The Netherlands.

Stiller, F. S.J. and R.P. Yadav. 1978. *Planning for People*. Human resources development centre, Kathmandu.

T. Prasad and Anand Verdhan. 1990. Management of conjunctive irrigation in alluvial regions-issues and approaches. International conference on groundwater resource management, Bangkok, Thailand.

Tahal Consulting Engineers 1999. Bhairahawa Lumbini groundwater irrigation project, stage III. Implementation Completion Report.

Thapa, Ganesh. B. and M.W. Rosegrant. 1995. Projection and policy implementation of food supply and demand in Nepal to the year 2020. Research report series no. 30. Winrock International. Kathmandu.

Thapa, Shanker. 2000. *Historical study of agrarian relations in Nepal (1846-1951)*. Adroit Publishers. Delhi.

Tillson, David. 1985. *Hydro-geological technical assistance to the agriculture for the Kingdom of Nepal.*

Uphoff, Norman. 1986. *Getting the process right: improving water management with farmer organisation and participation*. Ithaca: Cornell University.

Uprety, S.R. 1989. Shallow groundwater investigations in the Tarai Rupandehi District, UNDP. Project NEP/86/025. Nepal.

Utah Water Resources. 2005. Definition of conjunctive use from http://www.water.utah.gov/waterplan/uwrpff/Glossary.htm

Vincent, Linden and Peter Dempsey. 1991. Conjunctive water use for irrigation: good theory, poor practice. ODI Irrigation Management Network Paper 4, London, UK.

Vincent, Linden F. 1997. *Irrigation as a technology, irrigation as a resource: a sociotechnical approach to irrigation.* Inaugural speech. Wageningen, The Netherlands: Wageningen Agricultural University.

Vincent, Linden F. 1997. Irrigation as technology, irrigation as resource: hill irrigation and natural resource systems. *In People and participation in sustainable development: understanding the dynamics of natural resource systems*, eds. Ganesh Shivakoti, George Varughese, Elinor Ostrom, Ashutosh Shukla and Ganesh Thapa: 39-51. Proceedings of an international conference held at the Institute of Agriculture and Animal Science, Rampur, Chitwan, Nepal, 17-21 March, 1996. Bloomington, Indiana and Rampur, Chitwan, Nepal.

Vincent, Linden F. 2001. Struggles at the social interface: developing sociotechnical research in irrigation and water management. In *Resonances and dissonances in development. actors, networks and cultural repertoires*, eds. P, Hebbinck, and Gerard Verschoor. The Netherlands. Royal van Grocum : 65-81.

Vincent, Linden F. 2005. Science, technology, and agency in the development of drought prone areas: a cognitive history of drought and scarcity. Phd thesis. Development policy and practice, Technology Faculty, Open University, UK.

Vincent. Linden F. 1994. Parameters of doubt: prospects for groundwater assessment to help farmers in hard rock areas of South India in *Farmer management of groundwater irrigation in Asia*, eds. M.D.C. Abhayaratna, Douglas Vermillion and Johnson S. Perry: 47-53. Selected papers from a South Asian Regional Workshop. Colombo, Sri Lanka. International Irrigation Management Institute.

Vokes, R. 2001. Sourced at http://www.cgisp.org.np/sub.html

Wahaj, Robina. 2001. Farmers actions and improvements in irrigation performance below the Mogha: How farmers manage water scarcity and abundance in a large irrigation system in South-Eastern Punjab, Pakistan. PhD thesis. Wageningen University. The Netherlands

Water Resqurce Strategy Plan, 1997. WB/ JF. Kathmandu.

Yadav, Ram P. 1999. Land tenure situation in Nepal. Series 5.

Policy Outlook. Winrock International, Kathmandu, Nepal.

Yin, R. K. 1984. *Case study research: design and methods*. California. Sage Publications.

Yoder, Robert D. 1986. The performance of farmer-managed irrigation systems in the Hills of Nepal. Ph.d dissertation. Cornell University.

Yoder, Robert. 1994. Organization and management by farmers in the Chhattis Mauja irrigation system, Nepal. International Irrigation Management Institute. Research Paper No. 11. Colombo, Sri Lanka.

Zaag, Pieter van der. 1992. *Chicanery at the canal: changing practices in irrigation management in western Mexico*. PhD. Dissertation. Latin America Studies no. 65. Amsterdam: Centre for Latin American Research and Documentation.

Zwarteveen, Margreet and Nita Neupane. 1996. Free-riders or victims: women's nonparticipation in irrigation management in Nepal's Chhattis Mauja Irrigation Scheme. IIMI research report 7. Colombo, Sri Lanka: IIMI.